T0205559

Lecture Notes in Electrical Engineering

Volume 811

The book series *Lecture Notes in Electrical Engineering* (LNEE) publishes the latest developments in Electrical Engineering - quickly, informally and in high quality. While original research reported in proceedings and monographs has traditionally formed the core of LNEE, we also encourage authors to submit books devoted to supporting student education and professional training in the various fields and applications areas of electrical engineering. The series cover classical and emerging topics concerning:

- Communication Engineering, Information Theory and Networks
- Electronics Engineering and Microelectronics
- Signal, Image and Speech Processing
- Wireless and Mobile Communication
- Circuits and Systems
- Energy Systems, Power Electronics and Electrical Machines
- Electro-optical Engineering
- Instrumentation Engineering
- Avionics Engineering
- Control Systems
- Internet-of-Things and Cybersecurity
- Biomedical Devices, MEMS and NEMS

For general information about this book series, comments or suggestions, please contact leontina.dicecco@springer.com.

To submit a proposal or request further information, please contact the Publishing Editor in your country:

China

Jasmine Dou, Editor (jasmine.dou@springer.com)

India, Japan, Rest of Asia

Swati Meherishi, Editorial Director (Swati.Meherishi@springer.com)

Southeast Asia, Australia, New Zealand

Ramesh Nath Premnath, Editor (ramesh.premnath@springernature.com)

USA, Canada:

Michael Luby, Senior Editor (michael.luby@springer.com)

All other Countries:

Leontina Di Cecco, Senior Editor (leontina.dicecco@springer.com)

** This series is indexed by EI Compendex and Scopus databases. **

More information about this series at https://link.springer.com/bookseries/7818

Biswajit Mishra · Jimson Mathew ·
Priyadarsan Patra
Editors

Artificial Intelligence Driven Circuits and Systems

Select Proceedings of ISED 2021

 Springer

Editors
Biswajit Mishra
ULP-IC Lab
DAIICT
Gandhinagar, Gujarat, India

Jimson Mathew
Department of Computer Science
and Engineering
Indian Institute of Technology Patna
Patna, Bihar, India

Priyadarsan Patra
Vice-Chancellor's Office and School
of Computing
DIT University
Dehradun, Uttarakhand, India

ISSN 1876-1100   ISSN 1876-1119 (electronic)
Lecture Notes in Electrical Engineering
ISBN 978-981-16-6942-2   ISBN 978-981-16-6940-8 (eBook)
https://doi.org/10.1007/978-981-16-6940-8

This Springer imprint is published by the registered company Springer Nature Singapore Pte Ltd.
The registered company address is: 152 Beach Road, #21-01/04 Gateway East, Singapore 189721,
Singapore

Contents

About the Editors

Biswajit Mishra is a Professor and Head of ULP-IC Lab at Dhirubhai Ambani Institute of Information and Communication Technology (DA-IICT), India. He received his B.E. in Electronics and Communication Engineering from the National Institute of Technology, Jaipur in 1996. Dr. Mishra received his M.S. and Ph.D. degrees in Electronic Engineering from the University of Southampton, U.K., in 2003 and 2007, respectively. In 2007, he joined the University of Southampton as a Research Fellow and from 2010 to 2013 he was a Senior Scientist at ESPLAB-EPFL-Switzerland, working on ultra-low power and subthreshold design methodology and energy harvesting electronic circuits. His research interests include battery-less electronics, ultra-low-power circuits, sub-threshold design methodology, and system implementation for BSN and WSN.

Jimson Mathew is an Associate Professor and Head of the Department in the Computer Science and Engineering, Indian Institute of Technology Patna, India. He did his Masters in computer engineering from Nanyang Technological University, Singapore, and a Ph.D. degree in computer engineering from the University of Bristol, U.K. He has held positions with the Centre for Wireless Communications, the National University of Singapore, Bell Laboratories Research Lucent Technologies North Ryde, Australia, Royal Institute of Technology KTH, Stockholm, Sweden, and Department of Computer Science, University of Bristol, UK. Dr. Mathew is a Senior Member of IEEE and has also served as a Guest Editor for ACM TECS. His research interests include fault-tolerant computing, computer vision, machine learning, and IoT Systems.

Prof. (Dr) Priyadarsan Patra is the Pro Vice-Chancellor and Distinguished Professor of DIT University, India.He earlier served as a Chief Architect and Principal Scientist of the R&D divisions of Intel Corp (USA) and as the Dean of Computer Sciences at UPES University and Dean of Research at Xavier University. Over the years, he created and led world-class research and development for Servers, Systems-on-Chip, and smart devices. His research interests include Computer and Systems Architecture, Applications of Machine Learning, Big Data & Cloud Computing, and the Design of Low-power and Intelligent systems. Prof. Patra had developed an architecture for building Single Flux Quantum Circuits and adiabatic circuits for his dissertation.

Elected Senior Member of both the IEEE and the ACM and an IE Fellow, Prof. Patra is the Founding Chair of the global IEEE System Validation and Debug Technology Committee and serves on several international Boards of Advisors. Dr. Patra holds a Ph.D. degree in Computer Sciences from the University of Texas at Austin, an M.S. from the University of Massachusetts at Amherst, and a B.Engg from the Indian Institute of Science, Bangalore.

An Efficient and Affordable R-Pi Based Cardiac Disease Detection System

Neha Arora⑩ and **Biswajit Mishra**⑩

Abstract This paper proposes a Raspberry Pi (R-Pi) based system to automatically detect and classify most of the atrial and ventricular cardiac diseases. The system provides a necessary solution for resource constrained regions to timely detect fatal cardiac conditions. The R-Pi receives the user specific information from the ECG application specifically developed for the smart mobile phones and 2-Lead ECG data from the ECG sensors connected with limb electrodes and the Arduino Nano board. The ECG data along with the user specific information is further processed for lead separation, ECG feature extraction, and disease classification. The obtained results are sent to the doctor via email. Along with the system development, this work proposes various algorithms for feature points detection and disease classifications.

Keywords ECG · R-Pi · Disease classification · Feature extraction

1 Introduction

ECG recordings are frequently used to detect Arrhythmias that are relatively quiet in the early stages. However, they may provide valuable knowledge about an individual's fitness and aid in the detection of underlying heart anomalies and the timely detection can prove to be life saving. While not all of arrhythmias are permanent or necessitate medical treatment, they can signal the onset of serious heart diseases. An ECG depicts the electrical activity of the heart and provides a large amount of information on the functionality of the heart required for the proper diagnosis of

Supported by DST/SERB- CRG, Govt. of India, Research fund Ref: CRG/2019/004747.

N. Arora · B. Mishra (✉)
Dhirubhai Ambani Institute of Information and Communication Technology (DA-IICT), Gandhinagar, Gujarat, India
e-mail: biswajit_mishra@daiict.ac.in

N. Arora
e-mail: 201721007@daiict.ac.in

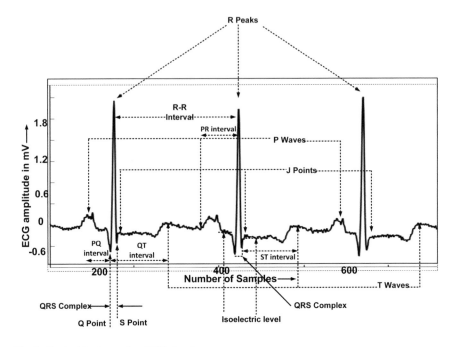

Fig. 1 Typical features of an ECG signal

various diseases. As illustrated in Fig. 1, a typical ECG comprises segments, waves, points, and QRS complex. These include P wave, Q point, R peak, S point, and T wave.

Cardiovascular diseases are one of the major causes of death globally [1]. In India, the average death rate due to cardiovascular diseases (CVDs) is significantly higher than the global average [2]. Due to the lack of primary health care in India, it is quite challenging to perform regular checkups in the hospital environments. It is even more troublesome for rural areas, where the availability of clinics and medical experts is comparatively low [3]. Hence, there is a significant need for automated, low-cost physiological testing systems that are simple to use, reliable, and can be used at home settings.

The development of a portable ECG monitoring system has been a topic of investigation over the past few years. In [4], a quasi real-time method was developed for separating out ventricular ectopic beats based on QRS template matching and R-R intervals assessment from both supraventricular and paced beats of the ECG. However, in order to operate, they required human assistance. In [5], ECG signal acquisition and classification of signals into three categories namely normal, Atrial Fibrillation, and Myocardial Infarction were performed on an ARM processor using wavelet analysis for feature extraction and support vector machine for the classification. A device that measured an ECG signal using an e-health sensor shield attached to an R-Pi was demonstrated in [6]. Further, MATLAB was used to analyze the heart rate variability based on ECG data in order to obtain the values. In [7], ECG sensors gathered patient physiological data, which was then analyzed by an Arduino micro-controller. LabVIEW software was installed on the doctor's phone, which accessed the patient's health details via email.

Our proposed system overcomes the dependency on offline processing by deploying the processing and disease detection algorithms on R-Pi. Equal to the size of a credit card, R-Pi operates as if it were a regular machine at a lower cost without any significant trade-off. Our system provides the results for most of the atrial and ventricular cardiac diseases such as Arrhythmias, Atrial Fibrillation, Hypocalcemia, Hypercalcemia, AV Blocks, and Myocardial infarction. Due to the Internet enabled and compact size of the system, it is easily deployed even at remote locations.

2 Proposed System Architecture

The proposed system architecture is shown in Fig. 2. To make the system more flexible to home environments, four limb reusable ECG clamp electrodes are used. The four electrodes, namely, RA (right arm), LA (left arm), LL (left leg), RL (right leg) are paired into two different leads with a pairing of RL, RA, and LA in Lead I and RL, RA, and LL in Lead II. These two leads are fed into two different AD8232 sensors through a 3.5 mm jack present on the board. The analog output from the sensors is fed into Arduino nano as it has the inbuilt ADC to convert the Analog ECG data to discrete sampled one and it is transmitted to R-Pi at a baud rate of 19200. Our system operates at a sampling frequency 184 Hz.

Additionally, our system considers specific criteria to assess the individual's preliminary health status. These factors include the individual's age, gender, and smoking status through an Android Mobile Application.

2.1 Android Mobile Application for User Specific Information

To develop an android mobile application, we have used the Message Queuing Telemetry Transport (MQTT) protocol. It is a lightweight, publish-subscribe network protocol that transports messages between different devices. The advantage of using this protocol is bandwidth efficiency and power.

The central communication point in the MQTT is the broker (R-Pi), which is in charge of dispatching all messages between senders (patients) and the rightful receivers (doctors). To obtain the information, the first step is to run a python script on the R-Pi to assign the status of R-Pi as a broker. To connect the android phone with R-Pi, we need to enter the IP address of the R-Pi on the mobile phone. After forming the connection between the two, the user specific details are filled and submitted. The

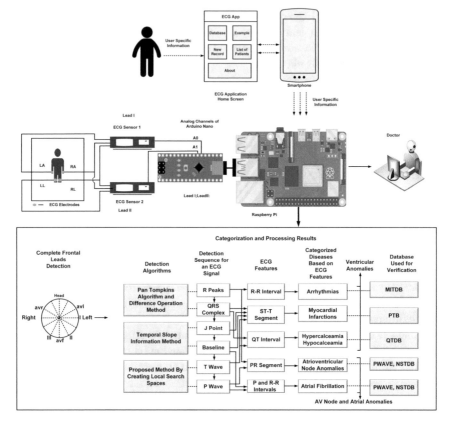

Fig. 2 Proposed system architecture

details filled will be stored in R-Pi in the text format. Figure 3 shows the screenshots of the developed android application and the proposed system architecture.

After receiving the user specific information the R-Pi starts storing the Lead I and Lead II ECG signals. Lead I and Lead II signals are utilized to obtain remaining frontal leads that are processed to classify various atrial and ventricular diseases. The obtained results include the Heart Rate Variability analysis of ECG and the summarized report on Arrhythmias, Atrial Fibrillation, PR anomalies, QT anomalies, and ST-T deviations.

3 ECG Signal Processing Algorithm

After determining all the frontal leads, the leads are processed for R peaks, QRS Complex, J point, baseline, P, and T waves of the ECG signal.

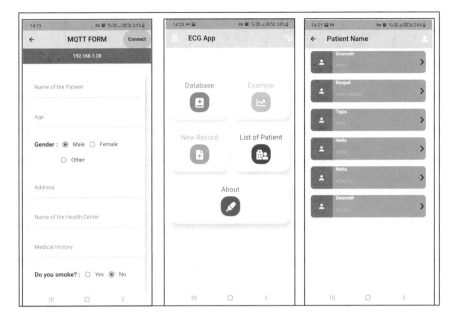

Fig. 3 Home pages of developed ECG application

3.1 QRS Complex, J Point and Baseline Detection

For detecting the R peaks, the widely accepted Pan Tompkins algorithm [8, 9] has been used. The QRS complex has been detected by the difference operation method stated in [10]. Detection of J point is done by creating a search space between $t_R +$ 20 ms to $t_R + 100$ ms duration, where t_R represents the location of the corresponding R peak. Three consecutive points with a slope less than or equal to 2.5 μV/sec are obtained and if the above-stated condition is satisfied, the midpoint is considered as the J point. Otherwise, 60 ms is added to the S point to be the J point [11]. The detection of the baseline of the signal is considered within a window spanning from $t_R - 100$ ms to $t_R - 40$ ms. The window is further searched for 20 ms with an average minimum slope value. The selected duration's average amplitude is considered as the baseline level of the signal. Search space for J point and Baseline is shown in Fig. 4. Baseline detection plays an important role in MI detection.

3.2 Proposed T Wave and P Wave Detection Algorithms

The T wave may be biphasic or monophasic in nature in one or the other leads and are necessary for detecting the cardiac anomalies such as MI, hypocalcemia, and hypercalcemia. To obtain T waves that may be biphasic or monophasic in nature,

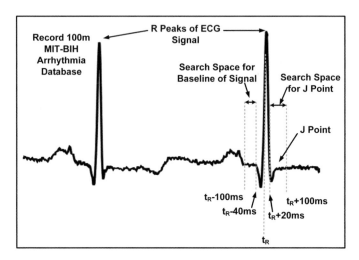

Fig. 4 J point and baseline detection of an ECG signal

we propose an algorithm that makes use of QRS complex, J point, and baseline of the signal as inputs with the ECG signal. This information is vital to create accurate search spaces for the probable T waves. The algorithm begins with the local maxima in the probable search spaces. During the course of experimentation, we observed that the signal is mainly affected by the baseline wandering and subtraction of adaptive baseline levels from the signal leads to improvement in results.

The ECG signal that is preprocessed with 0.5–10 Hz bandpass filter optimizes the T wave energy. Search space of 200 ms duration is created starting with the $Start_T$ point. t_J represents the J point location and RR(i) represents the corresponding RR interval. To make the search space adaptive of baseline changes and other drifts present in the signal, the baseline level has been subtracted from the search space. Further, we have determined maximum and minimum values in the adaptive search space with the respective locations. The proposed algorithm for T wave detection is shown in Fig. 5. Following this, the absolute maximum and minimum values are compared to obtain the biphasic or monophasic T wave location.

The detection algorithm of the P wave consists of preprocessing band pass filter to optimize the P wave frequencies and a stage to fix the QRS amplitude levels. It is done so that the QRS signal does not interfere in the detection of the P wave. After the preprocessing stage, we categorized the RR segments based on the Premature Ventricular Contractions (PVC) conditions. If the signal belongs to the PVC class then, the P wave detection for that particular case is not possible. Otherwise, a temporal search space for P wave as mentioned in [12] is considered. However, the mentioned literature utilized the phasor transform of the ECG signal, whereas we are processing the signal in the time domain.

The maximum's location in the search space is further compared with the previous cycle's T wave location. If the T wave location (t_T) and the P wave location (t_P) is

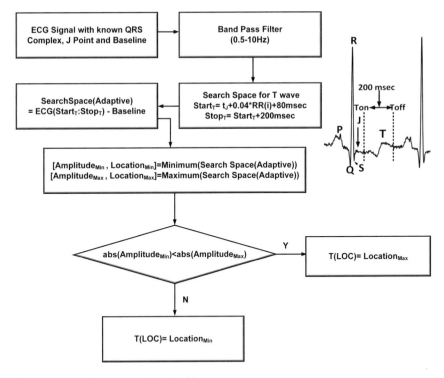

Fig. 5 Proposed T wave detection algorithm

same for more than 40% (value is chosen based on our observation during the experiments) times of the total length of T waves, it is assumed that the probable location for P waves is coinciding with the T wave locations. Therefore, the search space for P wave is shifted and the maximum amplitude in the selected duration is assumed to be the probable P wave location. Further, P wave amplitude ($Amplitude_P$) is compared with 1% value of R peak amplitude ($Amplitude_R$); if the condition satisfies, the P wave peak is considered as the valid P peak. Finally, t_P is searched for the local maximum in the 40 ms region for obtaining the final P wave location. The proposed algorithm for P wave detection is shown in Fig. 6.

3.3 Disease Classification

After determining all the feature points such as P, Q, R, S, and T waves, various features such as R-R interval, ST-T segment, QT interval, PR interval, P wave, and R-R features for cardiac diseases have been detected. Based on the R-R intervals of the Lead II signal, various arrhythmic conditions, namely Tachycardia, Bradycardia, R on

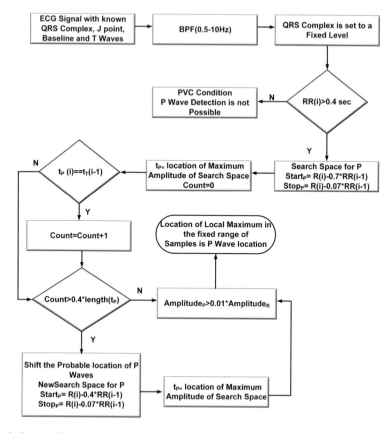

Fig. 6 Proposed P wave detection algorithm

T Condition, Asystole, Bigeminy, Trigeminy, Ventricular condition, and Interpolated premature ventricular conditions can be detected [13].

Similarly, for detecting and localizing four types of Myocardial Infarctions, viz., Anterior, Inferior, Inferolateral, and lateral cases an algorithm based on 6 Lead ECG signal (Frontal Leads) are discussed in [14].

QT interval is the time starting from the QRS complex to the end of the T wave. It represents the total time for activation of the ventricles and recovery to the normal resting state [15]. To determine the T_{off} point, a search space of 80 ms duration has been created that starts from $Tpeak + 40$ ms. For the given interval, we have determined the 16 ms window with the minimum slope value by using a moving window method and the starting point of that duration is considered as T_{off} point. QT interval is dependent upon the RR interval of the person. Hence, the corrected QT interval is used as a feature instead of directly using the QT duration. The corrected QT duration is given by Bazett's formula [15] as follows:

$$QT_c = \frac{QT}{\sqrt{RR}} \qquad (1)$$

The upper limit of QT_c duration is 460 ms. Similarly, the lower limit 390 ms [15]. If the values of the QT_c interval are beyond the threshold values, then it is considered as prolongation of QT duration (Hypocalcemia) or shortening of QT duration (Hypercalcemia).

For the detection of Atrial Fibrillation, P wave and R-R features are utilized. For 1 min of data, we have considered the Interquartile range (IQR) of R-R intervals as the variability measure. The missed detection of P waves for large number of cycles also signifies the Atrial Fibrillation condition. If the IQR is above 50 sample points (@ 360 Hz Sampling frequency) or in the case of missing P waves, it has been considered a probable case of Atrial Fibrillation.

The PR interval measurement is crucial for the detection of AV node anomalies. Ideally, the PR interval must exist between the range of 120 ms to 200 ms. The PR interval has been calculated for Lead II of standard 12 Lead ECG. If the values do not exist in the specified range, various AV node conduction anomalies such as AV block, AV nodal rhythm may exist.

4 Results and Discussion

For the validation of proposed algorithms, we have utilized various standard databases available on Physionet [16]. For example, for the detection of QRS complexes and arrhythmias, we have used MIT-BIH Arrhythmia Database (MITDB) [17]. Similarly, other databases such as PTB database (PTB) [18], QT database (QTDB) [19], MIT-BIH database with P wave annotations (PWAVE) have been utilized for specific features as shown in Fig. 2. Table 1 shows the QRS detection evaluation results at MIT-BIH arrhythmia database. The efficiency metrics are typically defined in terms of True Positives (TP), False Positives (FP), False Negatives (FN), and True Negative (TN) values. Based on these values, we have measured the Sensitivity (Se), Specificity (Sp), False Detection Rate (FDR), Positive Predictivity (PPV), etc. to validate our proposed algorithms.

Based on the detected R peaks, R-R interval has been utilized to detect various arrhythmias. A detailed discussion regarding the same can be found in our previous work [13]. Table 2 shows the present and detected arrhythmias on MIT-BIH arrhythmia database.

Table 1 Overall QRS detection evaluation

Total annotated beats	Beats detected	Average PPV%	Average Se%	Average FDR%
109864	109391	99.29	99.49	1.29

Table 2 Results for arrhythmias detection

Record	Arrhythmia present	Arrhythmia detected
100	PVC	PVC, Interpolated PVC
101	APC	PVC
102	PVC	PVC
106	Tachycardia, PVC, Bigeminy	Tachycardia, PVC, Trigeminy
108	PVC, Interpolated PVC	PVC, Interpolated PVC
114	PVC, Ventricular Couplets	PVC, Asystole, Bradycardia, Trigeminy
119	Trigeminy, Bigeminy	Trigeminy, PVC, Interpolated PVC
200	Tachycardia, Bigeminy, PVC, APC	PVC, Trigeminy, Tachycardia
201	Trigeminy	Trigeminy
203	Trigeminy, Tachycardia, Multiform PVC	Tachycardia, Trigeminy, PVC, Interpolated PVC
207	Bigeminy, Tachycardia	Tachycardia, Trigeminy, PVC, Interpolated PVC
214	Trigeminy, PVC, Tachycardia	Trigeminy, Tachycardia, PVC, Interpolated PVC
221	Tachycardia, PVC	Tachycardia, PVC, Trigeminy
222	Bigeminy	Tachycardia, PVC
223	Trigeminy, Tachycardia, Bigeminy, PVC	Tachycardia, Trigeminy, PVC, Interpolated PVC
232	Bradycardia, Asystole	Bradycardia, Asystole, PVC
234	Tachycardia, PVC	Tachycardia, PVC

Further, T wave detection results for the proposed algorithm have been validated on QT database and are shown in Table 3. The database consists of 15 MITDB signals and provides the annotations for the normal beats. Table 3 shows the evaluation of the T wave detection algorithm, where the overall sensitivity of the proposed algorithm is found out to be 97.78%.

Similarly, the P wave detection utilizes the MIT-BIH arrhythmia database with P wave annotations. The database consists of 12 databases from MITDB that are subcategorized to 5 normal and 7 abnormal waveforms consisting of various pathology conditions. Records #100, #101, #103, #117, and #122 belong to the normal category and Records #106, #119, #207, #214, #222, #223, and #231 belong to the abnormal categories. Mean Se, PPV, and FDR for normal waveforms are found out to be 99.39%, 99.42%, and 1.13%, respectively. Similarly, for the abnormal waveforms, the respective values are 94.23%, 82.74%, and 25.6% so, we conclude that the abnormal pathologies of waveforms lead to missed detection of P waves. Table 4

Table 3 T wave detection results for QT database for MITDB signals

Record	Annotated beats	TP	FN	Se%
sel100	30	30	0	100
sel102	85	84	1	98.82
sel103	30	30	0	100
sel104	77	74	3	96.10
sel114	50	50	0	100
sel116	50	48	2	96
sel117	30	30	0	100
sel123	30	30	0	100
sel213	71	69	2	97.10
sel221	30	26	4	86.6
sel223	30	30	0	100
sel230	50	50	0	100
sel231	50	47	3	94
sel232	30	30	0	100
sel233	30	30	0	100
Total	673	658	15	97.78

Table 4 P wave detection results for MIT-BIH Arrhythmia database with P wave annotations

Record	TP	FP	FN	Se%	PPV%	FDR%
100	2245	23	9	99.60	98.99	1.4
101	1835	15	27	98.55	99.19	2.2
103	2065	15	16	99.23	99.23	1.4
117	1527	3	4	99.74	99.80	0.45
122	2469	2	4	99.84	99.92	0.24
			Mean	99.39	99.42	1.13
106	1491	528	13	99.14	73.85	35.9
119	1566	455	52	96.79	77.49	31.3
214	1765	382	234	88.29	82.21	30.8
222	1118	265	135	89.22	80.83	31.9
223	2048	418	48	97.71	83.05	22.2
231	1878	18	114	94.27	99.05	1.6
			mean	94.23	82.74	25.6

shows the P wave detection results for the normal and abnormal signals present in the PWAVE database. Based on the detected P, Q, R, S, and T feature points, we have also detected the PR duration and QT duration of the signal that can be used to detect various cardiac conditions such as Hypocalcemia, Hypercalcemia, and AV Blocks.

Table 5 Interquartile range for atrial fibrillation detection for MITDB 1 min dataset

	Record no.	IQR (R-R)		Record no.	IQR (R-R)
Normal cases	100	27	Atrial fibrillation cases	201	79
	101	33.25		202	90.75
	103	37.75		203	279.25
	117	35.5		210	101.50
	122	23.25		217	50
				219	149
				221	155.75
				222	100.50

Table 6 Performance evaluation of proposed MI detection system with and without user specific information (USI)

USI	TP	FP	TN	FN	PPV%	Se%	Sp%	Acc%
✓	123	7	45	23	94.6	84.2	86.5	85
X	123	22	30	23	84.8	84.2	57.6	77.27

In addition to the formerly mentioned diseases, Lead II data can also be used to detect the Atrial Fibrillation condition by combining the P wave and R-R features. Irregular R-R interval and missed P wave signify atrial fibrillation. For a smaller set of ECG data (1 min of data), interquartile range (IQR) provides better variations approximation compared to the standard deviations. We have also observed that in the case of atrial fibrillation, R-R interval distribution is skewed instead of Gaussian and IQR provides the range of central 50% of data and removes the outliers. Table 5 shows the IQR for normal cases and atrial fibrillation cases for *1 min* of duration.

Along with the above-mentioned diseases, we have utilized 2-Lead ECG data to detect the Myocardial Infarction condition. A detailed description of Myocardial Infarction detection algorithm can be found in [14]. To validate the proposed algorithm, we have utilized the PTB Database. The database provides the *15* lead signals, but we have utilized the lead I and lead II signals to validate the Myocardial Infarction detection algorithm. We have utilized certain person specific parameters such as age, gender, and smoking status to validate our algorithm along with the 2-Lead ECG signals. It adopts and separates Myocardial Infarcted signals out of Normal Sinus signals and the efficiency metrics for the proposed system with and without using the user specific information (USI) are shown in Table 6.

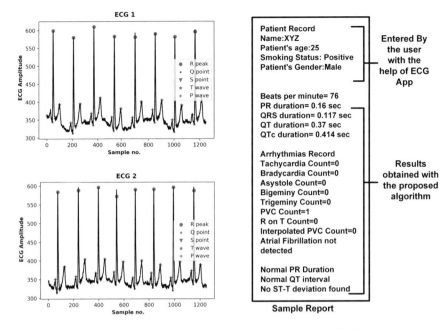

Fig. 7 Real-time P, QRS, and T detection; output sample report generated by the system and to be mailed to the doctor

4.1 Real-Time P, QRS, and T Point Detection

During real time, the system takes 60 s of data from the two leads as the input at a sampling frequency 184 Hz. However, for proper visualization of the detected points, a smaller dataset of 1250 samples, i.e., 6–7 s is used. Detected ECG outputs and generated output report are shown in Fig. 7. Along with this, our designed system is capable of sending the generated report to the doctor via email.

5 Conclusion and Future Work

Along with the system design, we quantified various feature points (P, QRS, and T) of the ECG signal and proposed various algorithms for ECG signal processing. The proposed algorithms are validated on the standard databases. QRS complex detection provides PPV, Se of 99.29% and 99.49% respectively. The proposed T wave detection approach based on local search spaces provides the detection Se of 97.78% for the manually annotated MITDB waveforms from the QTDB database. The proposed P wave detection scheme obtains the PPV, Se, and FDR of 99.42%, 99.39%, and 1.13%, respectively, for the normal cases. Similarly, for the abnormal cases, the values are 82.74%, 94.23%, and 25.6%. Myocardial Infarction detection is done on the available PTB database with PPV, Se, Sp, and Acc of 94.6%, 84.2%, 86.5%, and 85%, respectively. Our system provides efficiency of 100% in detecting Atrial Fibrillation. On the set of databases that we have considered, the effectiveness of both hardware and software, affordability and user friendly approach make this system an effective candidate for remote monitoring in resource constraint regions. The system's validation for cardiac patients is left as future work.

Acknowledgements We would like to acknowledge the contribution from Mr. Tejas Vasava and Mr. Smit Bhagat for the development of android mobile application. We would also like to acknowledge the contribution from Mr. Swaresh Phadke who helped us in porting the ECG signal processing algorithms to Raspberry Pi.

References

1. World Health Organization, Cardiovascular Diseases, World Health Organization Cardiovascular Diseases Fact Sheet available at: https://www.who.int/news-room/fact-sheets/detail/cardiovascular-diseases-(cvds) Last accessed 10 May (2021)
2. D. Prabhakaran, P. Jeemon, A. Roy, Cardiovascular diseases in India: current epidemiology and future directions. Circulation **133**(16), 1605–1620 (2016)
3. I. Silva, G.B. Moody, L. Celi, Improving the quality of ECGs collected using mobile phones: The Physionet/Computing in Cardiology Challenge 2011, in *2011 Computing in Cardiology* (IEEE, 2011), pp. 273–276
4. V. Krasteva, I. Jekova, QRS template matching for recognition of ventricular ectopic beats. Ann. Biomed. Eng. **35**(12), 2065–2076 (2007)
5. T. Jeon, B. Kim, M. Jeon, B.G. Lee, Implementation of a portable device for real-time ECG signal analysis. Biom. Eng. Online **13**(1), 1–13 (2014)
6. Ö. Yakut, S. Solak, E.M.İ.N.E. Bolat, Measuring ECG signal using e-health sensor platform (2014)
7. A. Abdullah, A. Ismael, A. Rashid, A. Abou-ElNour, M. Tarique, Real time wireless health monitoring application using mobile devices. Int. J. Comput. Netw. Commun. (IJCNC) **7**(3), 13–30 (2015)
8. J. Pan, W.J. Tompkins, A real-time QRS detection algorithm. IEEE Trans. Biomed. Eng. **3**, 230–236 (1985)
9. W.J. Tompkins, Biomedical Digital Signal, Processing: C-Language Examples and Laboratory Experiments for the IBM PC., *Englewood Cliffs* (PTR Prentice Hall, NJ, 1993)
10. Y.C. Yeh, W.J. Wang, QRS complexes detection for ECG signal: the difference operation method. Comput. Methods Programs in Biomed. **91**(3), 245–254 (2008)
11. S. Ansari, N. Farzaneh, M. Duda, K. Horan, H.B. Andersson, Z.D. Goldberger, B.K. Nallamothu, K. Najarian, A review of automated methods for detection of myocardial ischemia and infarction using electrocardiogram and electronic health records. IEEE Rev. Biomed. Eng. **10**, 264–298 (2017)
12. L. Maršánová, A. Němcová, R. Smíšek, M. Vítek, L. Smital, Advanced P wave detection in Ecg signals during pathology: evaluation in different arrhythmia contexts. Sci. Rep. **9**(1), 1–11 (2019)
13. B. Mishra, N. Arora, Y. Vora, Wearable ECG for real time complex P-QRS-T detection and classification of various arrhythmias, in *2019 11th International Conference on Communication Systems & Networks (COMSNETS)* (IEEE, 2019), pp. 870–875
14. N. Arora, B. Mishra, Characterization of a low cost, automated and field deployable 2-lead myocardial infarction detection system, in *2020 International Conference on Communication Systems & NETworkS (COMSNETS)* (IEEE, 2020), pp. 41–46

15. A. Houghton, D. Gray, *Making Sense of the ECG: A Hands-on Guide* (CRC Press, 2014)
16. A.L. Goldberger, L.A. Amaral, L. Glass, J.M. Hausdorff, P.C. Ivanov, R.G. Mark, J.E. Mietus, G.B. Moody, C.K. Peng, H.E. Stanley, PhysioBank, PhysioToolkit, and PhysioNet: components of a new research resource for complex physiologic signals. Circulation **101**(23), e215–e220 (2000)
17. G.B. Moody, R.G. Mark, The impact of the MIT-BIH arrhythmia database. IEEE Eng. Med. Biol. Mag. **20**(3), 45–50 (2001)
18. R. Bousseljot, D. Kreiseler, A. Schnabel, Nutzung der EKG-Signaldatenbank CARDIODAT der PTB über das Internet. Biomedizinische Technik/Biomed. Eng. **40**(s1), 317–318 (1995)
19. P. Laguna, R.G. Mark, A. Goldberg, G.B. Moody, A database for evaluation of algorithms for measurement of QT and other waveform intervals in the ECG, in *Computers in Cardiology 1997* (IEEE, 1997), pp. 673–676

Performance Evaluation of IoT Enabled Pedometer for Estrus Detection in Dairy Cows in India

Yasha Mehta and Biswajit Mishra

Abstract Insufficient herd fertility due to poor estrus detection significantly affects the dairy cattle sector. In India, the conventional way of detection relies heavily on herder's visual surveillance, which makes the quality of estrus detection highly subjective to herder's experience in identifying estrus signs. This method lacks automation and may not be effective in monitoring larger herds. Current devices used for cattle monitoring are imported, limited in services, and are often expensive. Continuous monitoring also suffers from poor battery life and is a deterrent for wide adoption. A low-cost system with long battery life and real-time monitoring could provide tremendous benefits to the Indian dairy industry. In this context, this paper presents our efforts to develop an IoT-based estrus monitoring and detection system that utilizes ICT technique to provide an effective map of the cattle reproduction phase and identify underlying health issues to determine the wellness of the animal specifically for Indian conditions. The system is developed to adapt to various deployment challenges such as frequent topological changes due to cattle's mobility, limited power source, and short transmission range of sensors. In this context, the performance of the proposed system has been evaluated through laboratory trials.

Keywords Estrus detection · Internet-of-things · Dairy · Activity monitoring · Pedometers

1 Introduction

Dairy industry is an essential part of the Indian economy. According to the latest research, the market size of dairy products in India was approximately INR 11,360 billion in 2020, with the milk being the major source of income [1]. Despite being

Y. Mehta (✉) · B. Mishra
VLSI and Embedded Systems Research Group, DA-IICT, Gandhinagar 382007, India
e-mail: yasha_mehta@daiict.ac.in

B. Mishra
e-mail: biswajit_mishra@daiict.ac.in

the largest producer, India exports relatively smaller volumes of dairy products and has managed to capture only a small share in the global dairy trade because, high milk production is due to larger bovine population [1]. Dairy farming forms an essential part of rural India, but milk production per animal is relatively low. Therefore, to increase profitability and satisfy the growing demands, it becomes necessary to increase the milk production of dairy cattle.

Insufficient cattle fertility poses a major threat that limits the reproductive efficiency of eligible cattle. It is due to poor detection of estrus [2] that negatively impacts milk production and affects fertility rates. This results in loss of livelihood of Indian farmers, leading to major economic losses. Therefore, effective detection of estrus is of prime importance as herd fertility, milk production, and farm profitability are strongly correlated.

During estrus, cattle show various symptoms like mucus discharge, sniffing, chin resting, mounting, restlessness, and increased movement. Various methods to detect estrus, reported in literature, require manual intervention and often cause discomfort to the animal. Owing to technological advancement, automated solutions for estrus detection are reported [3], which processes behavioral responses of the animals using state-of-the-art classifiers and machine learning algorithms. With the cattle showing various behavioral signs during estrus, high activity is considered as a prominent behavior and an effective indicator of estrus. Companies like Afimilk, DeLaval [4], Rumiwatch [5], ALT pedometer [6] have developed pedometer-based automated estrus detection systems that have become an essential part of herd management solutions. In India, the devices that are used are imported for large farms and are often expensive which are not affordable by small or medium farms. There have been a few experiments on Indian-specific studies [7] but they lacked automated data collection, and the device required frequent battery recharging. Therefore, in India, there is still no concrete solution available that addresses the concern of small- or medium-scale Indian farmers.

To facilitate Indian dairy farmers with efficient herd management tool, we propose precision livestock farming approaches combined with ICT to monitor and process activity behavior of cattle. The goal of this project is to develop a low-energy automated activity monitoring system for estrus detection that would help small- and medium-sized Indian farmers. In addition to providing remote real-time assessment of animal's behavior, the solution also offers easier integration of the existing dairy database [8] forming a part of a larger information network that would help to analyze regional trends. To the best of author's knowledge, this is the first complete solution developed indigenously and specifically for Indian farmers severely impaired by resource constraints.

2 Activity Monitoring System

2.1 General Description

Figure 1 shows the basic configuration of the proposed monitoring system. The major elements are a battery-powered pedometer sensor, sink nodes that can either be wall-powered or battery-powered and a cloud server. The pedometer is enclosed in a small package mounted on the animal leg to acquire activity information or steps taken. This recorded information is stored in the sensor buffer and is transmitted periodically to the sink nodes using a 2.4 GHz wireless link. The receivers or sink nodes are deployed for data aggregation at the server that contains algorithms and thresholds to interpret the data associated with animal_id. The results of the forecast are displayed on a front-end display to enable farmers and veterinarians to understand the status of the animal and help them to act on the next course.

2.2 Pedometer Design and Operation

The block diagram of pedometer shown in Fig. 1 is a custom hardware built that utilizes a micro-electro-mechanical system-based ultra-low power inertial measurement unit (IMU). To record the steps and activity of the animal, the system is initialized to step detection mode. A determined wake-up and sleep period is employed for energy conservation such that data transmission takes place only during active mode of the sensor and remains in power-saving mode for the rest of the time. Timers are used to

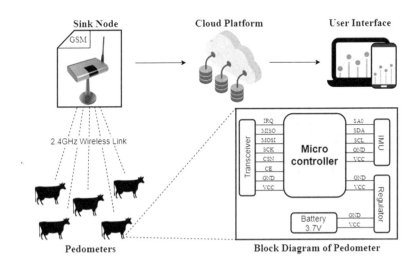

Fig. 1 Activity monitoring system

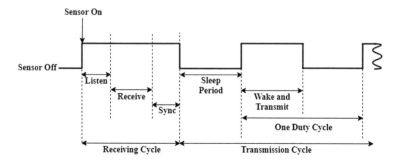

Fig. 2 Pedometer sensor operation

schedule the duty cycle of the pedometer. In order to provide a reference time for the timers to operate, when a pedometer tag is switched on as shown in Fig. 2, it remains in the wake state to receive the reference timestamp broadcasted by the synchronizer board. Upon its reception, the pedometer synchronizes its internal clock and start their transmission cycle. The pedometer tags calculate their sleep period according to their defined schedules and undergo a deep sleep mode for a calculated duration of time. Upon time-up, an interrupt routine is executed to wake the module for uplink. Figure 2 shows the complete operation of the pedometer tag.

In our developed system, the movement data is sampled at a rate 25 Hz and a step is detected only if more than 3 s of activity is observed to avoid misinterpretation of flexes with a step count. The on-board memory stores the acquired information during sleep periods and upon wake-up, the discrete data containing step counts and animal identification number(*animal_id*) is broadcasted. Since the receiver node is always awake, the wireless connection between the pedometer tags and the receiver (sink node) is established, and the data is successfully offloaded to the receiver.

2.3   Receiver Node Design and Operation

The receiver acts as a bridge between the pedometer tags/networks and the cloud server. It consists of a six-channel transceiver module operating over a frequency range of 2.400 GHz to 2.525 GHz with a separation of 1 MHz [9]. Each channel is designed to receive multiple pedometer's data simultaneously. They play a major role in expanding the system's capacity that is detailed in Sect. 3. The received data is uploaded to the server using a GSM module interfaced to the microcontroller using UART protocol. The communication between the GSM module and the database server utilizes HTTP protocol.

2.4 Cloud-Based IoT Platform

The cloud-based software environment utilizes the Microservices architecture as shown in Fig. 3 to classify the data and provide insights about the behavioral patterns of the animals. The microservice architecture is developed with the help of a Docker container. Each process is isolated into a different service that doesn't rely upon each other and has different functionality. They communicate with each other through the help of the REST API (Application Programming Interface) and GraphQL API, which both leverage the HTTP protocol. The three microservices in this architecture forming the back-end system are Hasura GraphQL Engine, PostgreSQL Database, and the back-end server based on Flask(Python). These services work in tandem to process the received raw data and it contains algorithms that records and processes the clean step entries for estrus detection and lameness detection. These microservices also communicate with the front-end UI based on NextJS which enables the server-side rendering of React Applications through the help of REST API and GraphQL API. Through the front-end UI, the processed data is displayed in the form of graphs, texts and important events (estrus status and health status) to be easily understandable by the veterinarians and farmers. Details of this architecture are discussed in [11].

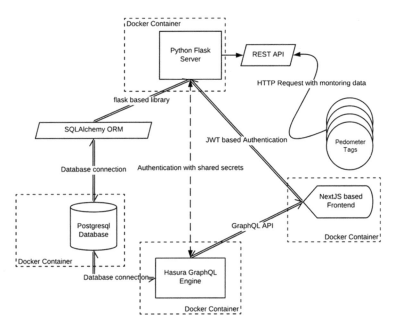

Fig. 3 Software architecture [10]

3 Performance Evaluation

It is important to ensure that the system functions reliably in monitoring the cattle's health along with estrus and present them remotely to the veterinarians and farmers at any time. Therefore, the functionality of each and every component of the system is to be evaluated and specific challenges owing to the implementation of sensor networks are to be addressed. The important parameters that need to be tested include Data Communication, Network Lifetime, Real-time assessment for identification of aberrant behavior and Scalability for large-scale deployment.

3.1 Data Communication

Network connectivity is important for data communication, but the problems such as cattle's mobility, changes in network topology, and radio interference from other animals significantly affect the system's performance. Therefore considering cattle's mobility, a one-hop star topology is employed at the network layer. Secondly, it may become possible that the radio transceivers at the sink node have a limited range of frequencies available to serve all the pedometers present within its vicinity. Thus, owing to the non-uniform distribution of the cattle, time division multiple access (TDMA) based channel access strategy is adopted such that each pedometer will have its own defined time-slot, which will alleviate data overlaps at the sink node.

To test for data communication, we set up six pedometers and three sink nodes in the laboratory. The pedometers were powered by a 3.7V LiPo battery and sink nodes were powered by wall socket plug-in power supply. The pedometers were placed randomly with one of the pedometer tied to a human leg for recording activity data, and the sink nodes were deployed 3 m apart such that at least one sink node was always present within the vicinity of the pedometer. The pedometers were programmed to have a duty cycle period, $D = 30$ min. The elapsed time 'D' is programmable and is a useful measure for system's scalability explained in Sect. 3.4.

At server's side, each time a new data is inserted, activity graph changes, and results of processing are displayed on the front-end in tandem. The graph in Fig. 4 maps the step data (analogous to the walking activity of the cow) recorded during a 7-day period. Based upon the results shown in Fig. 4, it is ensured that all the components of the monitoring system are working in accordance with the designed pathway of both hardware, software, and the network.

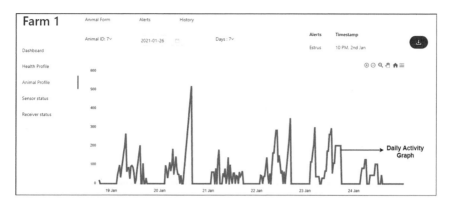

Fig. 4 Activity graphs at front-end

3.2 Network Lifetime

Network lifetime is an important attribute of any sensor networks and solely depends upon the sensor node's battery life. The life span of the sensor node relies on the current consumption of its components and is given by (1):

$$Lifetime = \frac{C_{capacity}}{I_{avg}}, \tag{1}$$

where $C_{capacity}$ is the battery capacity of the pedometer tag given in mAH and the average current, I_{avg} depends upon the power-saving strategies adopted. Since duty cycle based power-saving strategy is adopted, average current consumption can be calculated using (2).

$$I_{avg} = \frac{d_{active} * I_{active} + d_{sleep} * I_{sleep}}{D}, \quad \text{where, } D = d_{active} + d_{sleep}. \tag{2}$$

Here, I_{active} is the average current consumed for d_{active} seconds during active mode of the tag and I_{sleep} is the average current consumed for d_{sleep} seconds during the sleep mode.

Initially, the pedometer remains in the receiving mode; hence the current consumed during the receiving cycle is measured to be approximately 18.6 mA as shown by the highest current spike in Fig. 5. The periodic spikes in Fig. 5 shows the current consumed during data transmission which is measured to be around 4.8 mA. In between subsequent transmissions, the module remains in a deep sleep mode to conserve energy thus, the sleep mode current consumption is measured to be around 120 μA.

In reference to Fig. 5, after synchronization the module continues to function in the transmission phase; therefore the pedometer's lifetime majorly depends on the average current consumed during the transmission cycle. Thus $I_{active} = 4.8$ mA and

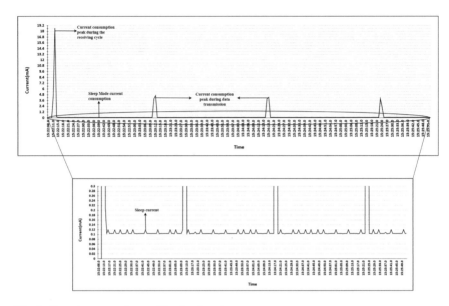

Fig. 5 Current consumption profile of pedometer sensor

$I_{sleep} = 120\,\mu A$. Now, with the duty cycle period $D = 30\,min$, the wake-up period, d_{active} is observed to be 4 s and sleep period, d_{sleep} is observed to be 1796 s. So, the average current calculated using (2) is around 0.1304 mA per duty cycle. The pedometer tags are powered by 3.7 V/1500 mAh LiPo battery, thus the lifetime of the pedometer tag is calculated to be around 2.63 years and based on lab experiments, it is observed to be approximately 2.84 years.

3.3 Real-Time Assessment for Identification of Aberrant Behavior

In order to determine the performance of the system in identifying estrus or possible ailment, actual activity values of the animal are compared to its baseline behavior in a particular time frame. The baseline behavior is the threshold values obtained by computing the average of the step counts plus twice the standard deviation within the similar time-interval of the preceding 10 days. Voting algorithm is used for the detection such that if the number of steps exceeds its respective threshold values for a defined number of times in a 3 h period, estrus is concluded while the lameness is detected if reduction of 5% or more is observed in daily activity when compared to the threshold values in a 5 h period. The daily activity mapped in 'dark' traces is compared against the 'gray' traces of baseline activity of the same animal to analyze the behavioral responses. The beginning of estrus is indicated with the increased

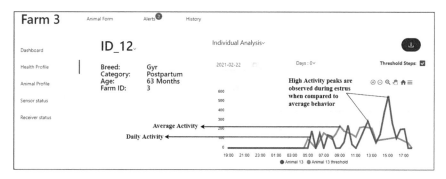

Fig. 6 Activity analysis of individual Animal—Estrus

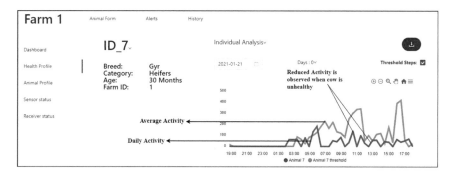

Fig. 7 Activity analysis of individual Animal—Lameness

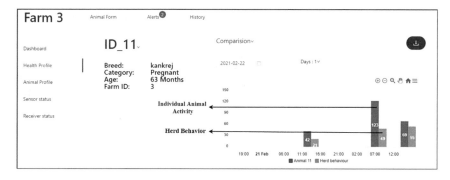

Fig. 8 Activity comparison against average herd behavior

activity as illustrated in Fig. 6 while the lameness is indicated with decreased activity as illustrated in Fig. 7. In addition, average herd behavior is also recorded which is used to compare an individual animal's behavior against the behavior of the herd it belongs, to determine its performance. Figure 8 shows particular animal's activity in 'dark' bars and average herd behavior in 'gray' bars.

It is to be noted from Figs. 6 and 7 that the activity analysis of two different animals is from two different farms, i.e., Farm 1 and Farm 3. This verifies the system's feature which helps the farmer in managing all his farms remotely from one single point. Also, a facility to download the recorded information, in the form of a pdf file is provided, which contains details of the animal of interest and activity trends associated with them. This enables the farmer to share the recorded data to the veterinarians to determine well-being of the animal or to the buyers to know the history of the animal before a purchase.

3.4 System Scalability

Reliability and scalability to large-scale deployment are of prime importance for large farms accommodating hundreds of cows. The sink nodes incorporated with multi-channel transceiver module play a major role in expanding system's capacity by exploiting all the available data pipes for data reception. The transceivers have six data channels [9] and are capable of receiving 'n' pedometers transmission simultaneously per time-slot where, $1 \leq n \leq 6$. Experiments were conducted with the six pipes enabled for reception such that $n = 6$. Each pedometer had its unique pipe address for the time-slot assigned to it as illustrated in Fig. 9, so it is certain that a packet will be successfully received on each transmission. For example, out of 1000 packets, 776 packets were received correctly. Thus the sink node's throughput efficiency was observed to be around 77.6%.

It should be noted that the transceiver operates over a frequency range from 2.400 to 2.525 GHz. With the separation of 1 MHz, 125 frequency channels are available for communication [9]. By utilizing different communication frequency, F_s for different sections 's', we can increase the capacity of the system for larger farms having tens of hundreds of cows. For implementing this, it was important to check for data interference between the adjacent frequency channels. To verify this, we set up one set of sink node and two pedometers communicating at $F_1 = 2.440$ GHz and another

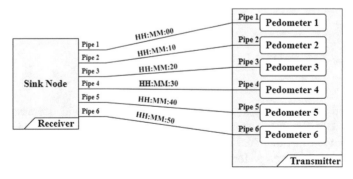

Fig. 9 Demonstration of multi-channel reception

Fig. 10 Setup to check for
data interference

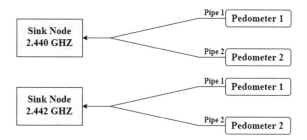

at $F_2 = 2.442\,\text{GHz}$ as shown in Fig. 10. The performance of a complete system
was examined for a month. No data interference was observed between adjacent
communication frequencies. Thus, it can be concluded that the system capacity to
accommodate number of cows denoted by 'C' depends upon the number of data
pipes (n), duty cycle period (D), and number of frequency channels (F) used and is
given as

$$C = n * D * F, \quad \text{where,} \quad F = \sum_{s=1}^{125} F_s. \tag{3}$$

Suppose, if $n = 6$, $D = 30\,\text{min}$ and $s = 2$, system can accommodate 360 cows.
Similarly if $n = 6$, $D = 60\,\text{min}$ and $s = 125$, the theoretical limit that the system
can accommodate is 45000 cows. Thus, it is ensured that the system can support
monitoring tens of hundreds of cows within one farm. With frequency reuse, more
number of cows is possible to be accommodated and is left as a future work.

4 Conclusion

In India, there are few farms/organizations that have adopted modern herd monitoring technologies. The available devices are imported and often expensive which are
not affordable by small- or medium-scale farmers. Since the opportunity to identify
a cow in estrus is relatively short, lasting a few hours in the 21-day cycle, with the
cows showing fewer signs of estrus and for a shorter duration, it becomes difficult
for farmers to identify estrus by only visual observations. Therefore, this research
aims to develop an activity-based estrus detection system with ultra-low energy consumption to be advantageous for small- or medium-sized Indian farmers in resource
constrained regions. The IoT-based software infrastructure will facilitate the farmers
with remote real-time tracking of the cattle and identify estrus triggers; giving them
a chance to prevent missing cycles. In addition, performance of the individual cattle
can be tracked and compared to the herd it belongs such that deviation from normal
will enable early diagnosis of underlying health issues. The research also facilitates

data integration with existing regional or national databases to form a part of larger information network. To make this system more usable, the proposed system has the potential to be extended for other applications such as lameness detection or standing time analysis, rumination monitoring and body temperature measurements that could be useful for dairy farmers.

Acknowledgements This work is a part of project (19/STDST/BSM/DFTEA) funded by Department of Science and Technology under S&T scheme. The authors acknowledge the efforts of interns Mohammed Shadab and Vedant Thakkar in developing the cloud platform for data analysis and visualization.

References

1. IMARC Group: Dairy Industry in India, *Edition: Market Size* (Procurement and Distribution, Growth, Prizes, Segments, Cooperatives, Private Dairies, 2021), p. 2021
2. P.L. Senger, The Estrus detection problem: new concepts, technologies, and possibilities. J. Dairy Sci. **77**(9), 2745–2753 (1994)
3. B. Sharma, D. Koundal, Cattle health monitoring system using wireless sensor network: a survey from innovation perspective. IET Wirel. Sens. Syst. **8**(4), 73–80 (2018)
4. P. Lovendahl, M.G.G. Chagunda, On the use of physical activity monitoring for estrus detection in dairy Cows. J. Dairy Sci. **93**(1), 249–259 (2008)
5. Alsaaod, M., Niederhauser, J.J., Beer, G., Zehner N., Schuepbach-Regula, G., Steiner, A.,: 'Development and validation of a novel pedometer algorithm to quantify extended characteristics of the locomotor behavior of dairy cows. J. Dairy Sci. **98**(9),(2015)
6. U. Brehme, U. Stollberg, R. Holz, T. Schleusener, ALT pedometer-new sensor-aided measurement system for improvement in estrus detection. Comput. Electron. Agric. **62**(1), 73–80 (2008)
7. S. Kerketta, T.K. Mohanty, M. Bhakat, A. Kumaresan, R. Baithalu, R. Gupta, A.K. Mohanty, M. Abdullah, S. Kar, V. Rao, A. Fahim, Moosense pedometer activity and Periestrual hormone profile in relation to Oestrus in crossbred cattle. Indian J. Animal Sci. **89**(12), 1338–1344 (2019)
8. NDDB Database. https://www.nddb.coop/services/sectoral/national-database
9. Nordic Semiconductor: NRF24L01 Single Chip 2.4GHz Transceiver Product Specification (2nd edn., 2007)
10. Cloud Services. https://www.digitalocean.com/
11. Software Architecture Description. https://medium.com/code-dementia/yoctosehns-software-architecture-cff81658351c
12. R. Firk, E. Stamer, W. Junge, J. Krieter, Automation of estrus detection in dairy cows: a review. Livest. Product. Sci. **75**(3), 219–232 (2002)
13. A.T. Peter, W.T.K. Bosu, Postpartum ovarian activity in dairy cows: correlation between behavioral estrus. Pedometer measurements and ovulations. Theriogenology **64**(8), 111–115 (1986)
14. J.B. Roelofs, F. Eerdenburg, N.M. Soede, B. Kemp, Pedometer readings for estrous detection and predictor for time of ovulation in dairy cattle. Theriogenology **64**(8), 1690–1703 (2005)
15. I. Andonovic, C. Michie, P. Cousin, A. Janati, C. Pham, M. Diop, Precision livestock farming technologies, in *2018 Global Internet of Things Summit (GIoTS)* (IEEE, 2018), pp. 1–5
16. K.H. Kwong, T.T. Wua, H.G. Goh, K. Sasloglou, B. Stephen, I. Glover, C. Shen, W. Du, C. Michie, I. Andonovic, Practical considerations for wireless sensor networks in cattle monitoring applications. Comput. Electron. Agric. **81**, 33–44 (2012)

Toward an Adversarial Model for Keystroke Authentication in Embedded Devices

Niranjan Hegde and Sriram Sankaran

Abstract The increase in computing and communication capabilities has enabled embedded devices to perform a wide variety of sensing and monitoring tasks in diverse domains. Keystroke authentication for embedded devices leverages the unique typing patterns of users toward creating behavioral profiles for identification. However, keystroke authentication can be bypassed by manipulating inputs thus causing machine learning models to operate on perturbed data resulting in higher error rates. In this paper, we propose to develop an adversarial model for keystroke authentication in embedded devices. The proposed model leverages Feature Importance Guided Attack (FIGA) [1] method to add perturbations in the inputs which in turn increases error rate thus allowing the attacker access to the system. Evaluation using RandomForest, XGBoost, and SVM models shows an average Error rate across all the users to be 60%, 55%, and 63%, respectively. Our analysis shows that it takes an average of 3.18 s for the attacker to generate samples per user and that the generated samples are realistic in nature. Finally, the computational complexity of our proposed model is O(n) which is comparable with existing adversarial models.

Keywords Adversarial model · Keystroke authentication · Embedded devices

1 Introduction

The rapid proliferation of embedded devices with advanced computing and communication capabilities has given rise to applications in diverse domains [2]. These devices range from tiny resource-constrained sensors to more powerful data centers typically known as Internet of Things which are capable of remote monitoring and response. The advent of machine learning for embedded devices is considered a "double-edged sword" which means that developing machine learning models can

N. Hegde · S. Sankaran (✉)
Center for Cybersecurity Systems and Networks, Amrita Vishwa Vidyapeetham,
Kollam, KL, India
e-mail: srirams@am.amrita.edu

© The Author(s), under exclusive license to Springer Nature Singapore Pte Ltd. 2022
B. Mishra et al. (eds.), *Artificial Intelligence Driven Circuits and Systems*,
Lecture Notes in Electrical Engineering 811,
https://doi.org/10.1007/978-981-16-6940-8_3

be used to both enable as well as disable security for embedded devices [3]. These scenarios are increasing prevalent given the increasing challenges of embedded applications coupled with the varying capabilities of attackers.

Passwords are typically used in embedded devices for authentication [4]. These devices are typically pre-configured with default passwords which can be guessed with ease and authentication process bypassed. Thus, passwords involving a combination of numbers, capital letters and special characters are required to increase the difficulty of password guessing. Despite these requirements, attackers are able to bypass them since the requirements are typically not followed.

2-Factor Authentication (2-FA) has increasingly become the de-facto standard for system requiring high-level security for systems such as banking. The idea behind 2-FA is to authenticate users based on what they know (password) and what they have (token) [5]. With the advancement in technology, biometric-based authentication has become a viable option. Authentication system may use information such as the retina eye scan and fingerprints for authentication. However using biometric form of authentication raises privacy concerns [6].

Keystroke-based authentication is considered a form of biometric authentication with low risk to the privacy of the user because the users are identified based on their typing patterns. It can also be combined with mouse dynamics to further enhance the authentication process. In a keystroke-based authentication, system stores the keystroke timing patterns of the users while typing. Based on their typing patterns, users are authenticated by the system. To facilitate keystroke authentication, system typically uses machine learning algorithms to classify the typing patterns of users.

However, keystroke-based authentication models are not secure since they can be bypassed to allow attackers access to the system. In particular, data or the model can be manipulated to cause perturbations thus causing the machine learning models to misclassify data. This emphasizes the need for modeling the adversarial behavior on keystroke authentication and analyzing its impact on embedded devices for varying number of users.

In this paper, we propose to develop an adversarial model for keystroke authentication in embedded devices. In particular, we leverage the FIGA [1] method to add perturbations in model inputs thus causing the machine learning models to misclassify data and further enable the attackers access to the system. We also compare the computational complexity of our proposed adversarial model against existing adversarial model. Evaluation using machine learning models such as RandomForest, XGBoost and SVM shows an average Error rate across all the users to be 60, 55 and 63% respectively. Our analysis shows that it takes an average of 3.18 s to generate samples for all users and the average change in generated samples across all users is 1 s which denotes the realistic nature of the samples. Finally the computational complexity of our proposed model is $O(n)$ which is comparable with existing adversarial models.

2 Related Work

There exists numerous forms of authentication for embedded devices such as authentication using Photo Response Non-Uniformity (PRNU) of a smartphone described by Nimmy et al. [7]. In our work, we focus on the existing literature on keystroke-based authentication systems and developing adversarial models.

Keystroke authentication: In [8], authors crowdsourced using Amazon's mechanical turk [9] where they collected typing patterns of different users. They selected the password ".Generation!8765" which users need to type for 5 times. They selected 3 features: Key-hold, Key-flight, and Key-latency. Key-hold refers to the time until the key is pressed. Key-flight refers to the time key is pressed and the next key is pressed. Key-latency refers to the time between leaving the key and next key is pressed. In their paper, they have set the threshold as 75% of the hold time and standard deviation for the feature should be within 2.5 standard deviations. With this approach, they had a false acceptable rate of 0.00% however false rejection rate was 13.89%.

In [10], authors used the dataset provided in [11] and evaluated various machine learning such as KNN, SVM Random Forest, and XGBoost. They found that XGBoost got the highest accuracy of 93.5%. Similarly, in [12], authors also showed that XGBoost was the preferred model for keystroke-based authentication model. Authors have obtained an accuracy of 90.91% using their approach. In our paper, we have launched the attacks against these machine learning models to demonstrate our approach.

In [13], authors used the dataset provided in [11] and developed an Extreme learning model (ECM) and Extreme learning model combined with evolving cluster model (ELM-ECM). Using the dataset, ECM achieved an accuracy of 89.27% whereas ELM-ECM obtained 86.97%. Since the accuracy is not higher than XGBoost, we did not launch the attack against this machine learning model as a part of our current work. In future work, we propose to consider these deep learning models as well.

Data Perturbation: Negi and Sharma [14] performed adversarial attack on keystroke authentication model where they took the average of the typing patterns and created an artificial sample. Further, k-means clustering was utilized to generate artificial samples per user. The authors found that they were able to compromise 50% of the users using their proposed mechanism. Negi et al. [15] proposed a novel algorithm K-mean++ which generates adversarial samples by iterating over the samples and passwords. They also show that this algorithm could be used to attack other forms of biometric-based authentication such as touchscreen swipes. They compromised 40–70% of the users by using varying degree of their attack strength.

Khan et al. [16] developed a mimicry attack on smartphones which guides attackers to provide inputs similar to the victims entering passwords. In particular, authors assumed that they have access to the victim's password and typing patterns after having captured them while typing. They also train attackers via help of a visual aids to enter the password similar to that of a victim. However, this paper does not focus on exploiting machine learning but instead on accurate mimicry of the victim's typing password once the attacker has learnt about the password.

Sun and Upadhyaya [17] created a master key using a large dataset where wolves and lambs were identified. Using lambs and wolves, they create a master key which is used to increase the Equal Error Rate (EER). The authors were able to increase EER by 22%. Their approach requires knowledge of the dataset to determine the lamb and wolves to attack the model.

In contrast to the existing approaches, we develop a novel adversarial model for keystroke authentication in mobile embedded devices using Feature Importance Guided Attack (FIGA) method.

3 Keystroke Authentication: Background

Figure 1 describes the working of a keystroke-based authentication model. First, users would register their typing patterns of their passwords. Further, the system would preferably ask the user to type in the same word a few times to account for any deviations. Then, the system would keep track of the time taken for pressing a key, releasing the pressed key, and total time until the next key is pressed after a key is released. Finally, the system would store the typing patterns of users and learn to differentiate among them.

The system for uniquely identifying the typing patterns of users can be developed using statistical methods and machine learning models. Once these models are able to differentiate among the users based on their typing patterns, they are further used for identification.

When the user tries to login to the system by typing in the password, it would first check for the validity of the password. If the password is successful, system would track the time taken by the user to press the key and latency between the keys. These

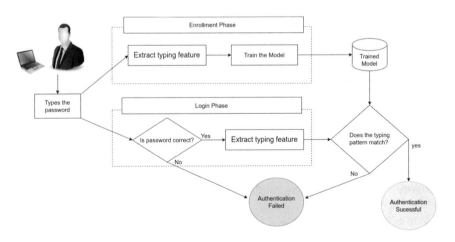

Fig. 1 Keystroke authentication model

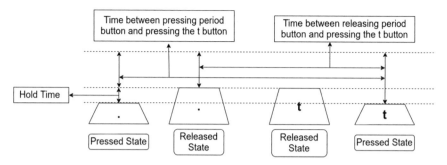

Fig. 2 Feature extracted from typing pattern

details would then be fed to the trained model to predict whether the typing pattern matches. If it matches, then the system would enable the login. On the other hand, authentication is rejected if password is invalid.

Figure 2 shows the information captured during the process of typing. Three feature values that are computed are the following:

- Hold time denotes the time between pressing the button and release.
- Time between pressing the first and the next button.
- Time elapsed after releasing the first button and pressing the next button.

For every adjacent key-pair in the password, above feature values are computed. In Fig. 2, we have assumed that the key-pair to the period button and character "t" button. If the next button to press is the character "e" button, then the key-pair would become "t" and "e" button. The feature values would be calculated again for this key-pair. It is necessary that the sum of hold time of the first button and time between releasing the first button and pressing the next button is equal to the time between pressing the first and the next button. Thus, feature values are extracted from the typing patterns which is then used for training the model.

4 Proposed Adversarial Model

In our proposed adversarial model, we have made the following assumptions:

- Attacker knows the password and typing patterns of the users.
- Attacker can provide generated samples as input.

Figure 3 contains the pictorial description of the proposed adversarial model. The attacker utilizes the testing samples to compute the importance of each feature and rank them accordingly. After performing feature ranking, attacker needs to select two parameters ϵ and n which are used to control the amount of perturbation that can be added in the samples. Finally, attackers use FIGA to generate artificial samples of typing patterns that are then used to attack the keystroke authentication model.

Fig. 3 Flowchart of
proposed adversarial model

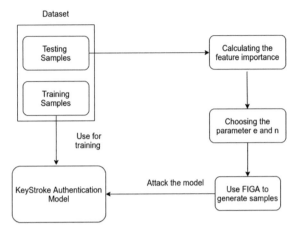

FIGA is modified such that the relationship between features as explained in
the earlier section is maintained thus ensuring realistic perturbations. The modified
samples when provided to the keystroke detection model cause the Error Rate to
increase thereby allowing the attacker to bypass the model and gain access to the
system.

Algorithm 1 Modified FIGA Algorithm

    Input: f, n, ϵ, x
    Output: x^*
1: $f_n \leftarrow f[n]$
2: $\epsilon \leftarrow \frac{\epsilon}{n} * sum(x_{features})$
3: **for** $i \leftarrow 0, n$ **do**
4:     $x[i] \leftarrow x[i] + (\epsilon * d[n])$
5:     **if** $x[i] < 0$ **then**
6:         $x[i] \leftarrow 0.01$
7:     **end if**
8: **end for**
9: **return** x^*

Dataset The CMU dataset [11] is being used in the paper. The dataset contains
the typing pattern of several subjects. Subjects typed the password ".tie5Roanl" in
the keyboard. Each row contains the typing information of a single repetition of the
password by a single subject. The feature set contains three types of features: the
hold time for a key, the time between pressing the next key when the current key is
released, and time between when the key and the next key is pressed.

Modified FIGA Algorithm Before the algorithm is utilized, feature importance
ranking of the feature is performed using any ranking method. In this paper, Gini-
impurity method is used for feature ranking. Once the feature has been ranked, the
direction of each feature is computed by taking the sign after subtracting the mean of

the feature values of the target users and those of the feature values of the rest of the users. The computed direction can be negative or positive through which direction of perturbation is known.

After computing the feature ranking and its direction, modified FIGA algorithm is applied. Inputs to Algorithm 1 are f (Rank and attack direction of the feature), x (samples to perturb) ϵ (the total percentage to modify) and n (number of feature to modify).

Step 1: Given n features from the list of ranked feature importance, additional features are added into the list to maintain the relationship. For example, if the feature to be modified is hold time of the key "e", then the feature where the time between the pressing of key "e" and the next key is also added into the list.

Step 2: In this step, sum of the feature vectors is computed and multiplied with ϵ. The resulting value denoting the total amount of perturbation is divided by n to effectively distribute the perturbation among the selected features.

Step 3: In this step, the resulting perturbation obtained from Step 2 is multiplied with the direction of the feature before being applied to the features. Perturbations are being added to only n features ranked by the feature importance ranking method.

Step 4: In this step, negative values in samples are replaced with the value of 10 milliseconds to ensure that samples generated are realistic in nature.

Step 5: In this step, we obtain the adversarial samples created by the modified FIGA algorithm which are then tested against machine learning models.

Experiment Setup: We use the Scikit-Learn library [18] running on a laptop machine with 16 GB RAM 8 core process for implementing the machine learning models. We utilize 80% of the dataset for training and 20% for testing the trained model. The adversarial samples were created from the testing dataset. The value of ϵ and n is set to 0.01 and 3 so that the perturbations added are minimal and optimal.

We select Random Forest Classifier, XGBoost Classifier, and Support Vector Machine (SVM) for Keystroke Authentication. The rationale behind selecting the above models is that they provide better performance compared to other approaches.

5 Evaluation

In this section, we evaluate our proposed adversarial model using Random Forest classifier, XGBoost classifier, and Support Vector Machine. Further we perform the analysis of the overhead and the crafted adversarial samples. Finally, we compare the computational complexity of our proposed model against existing adversarial models.

Metric: The goal of our proposed model is to increase the classification error in keystroke authentication model. The chances of the attacker authenticating as another user increases with the increase in classification error. Therefore, we use Error Rate as our metric to evaluate the performance of keystroke authentication model and our proposed adversarial model. Error Rate is computed by dividing the sum of False

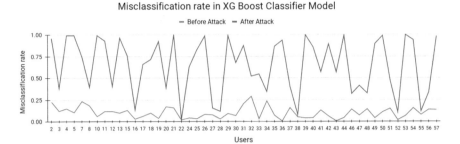

Fig. 4 Error rate per subject before and after attack (XGBoost Classifier)

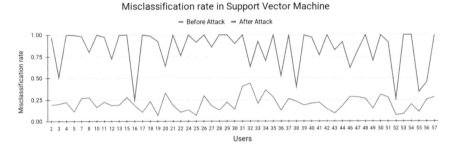

Fig. 5 Error rate per subject before and after attack (Support Vector Machine)

negatives and False positives by the total number of predictions. Error Rate can be expressed using the following equation:

$$Error\,Rate = \frac{False\,Positives + False\,Negatives}{Total\,Number\,of\,Prediction} \quad (1)$$

Results: We test our modified FIGA algorithm against machine learning models such as RandomForest, XGBoost, and Support Vector Machines (SVM). Using the RandomForest model, Error rate for a few users such as user 53, user 42, user 39, etc. is 99% as shown in Fig. 6. Similar results can be seen for SVM (Fig. 5) and XGBoost models (Fig. 4) as the error rate for the same set of users was over 95%. This shows that these users are a good target for the attacker to use in order to bypass the system. Thus, making minimal changes to the typing patterns of any of these users, the attacker can masquerade as another user for authentication.

On the other hand, there are few users whose typing patterns are unique and resilient to manipulation such that it does not cause a significant increase in error rate. From Figs. 4, 5 and 6, it is evident that there are users whose error rates are less than others. This means that the attacker masquerading as another user for authentication would likely be unsuccessful.

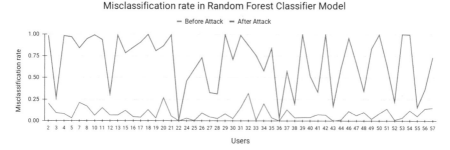

Fig. 6 Error rate per subject before and after attack (Random Forest Classifier)

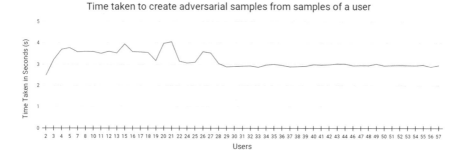

Fig. 7 Time taken to create adversarial samples from all samples of a user

5.1 Overhead Analysis

Figure 7 contains the results for overhead analysis of the proposed adversarial model. In particular, we estimate the time it takes to generate adversarial samples from all the user samples. On an average, it takes 3.18 s to generate adversarial samples across all users. From the figure, we observe that generating samples for the first 27 users requires more than 3.5 s, but incurs 2.9 s for the remaining users.

This difference can be attributed to the computation of feature importance using a Random Forest Model. This model is first fitted onto the samples provided by the attacker. The model uses Gini-Impurity to compute importance values for each feature. For users before 27, Random Forest Model requires more time to fit itself onto the samples.

5.2 Analysis of Crafted Adversarial Samples

In our work, we use modified version of FIGA to generate samples that increase the error rate in the keystroke authentication model. This allows the attacker to

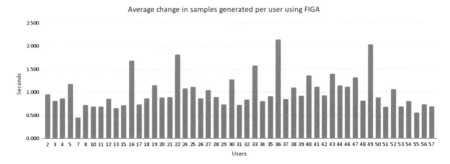

Fig. 8 Time taken to create adversarial samples from samples of all user

authenticate himself as another user. In this section, we examine the changes in generated samples to evaluate their realistic nature.

Figure 8 shows the average change per user in samples generated using modified version of FIGA. Average change per user is calculated by taking the absolute mean of the differences in original samples and generated samples using modified version of FIGA. We observed the maximum and minimum average change to be 2.151 s and 0.459 s respectively. This shows that the samples generated using modified version of FIGA are realistic. An attacker can use these samples to attack a keystroke authentication model.

The overall average change across all users is 1 s. In Fig. 8, typing samples generated from users 16,22,36 and 49 show an average change greater than 1.5 s. This can be attributed to the computation of the average change per sample by the modified version of FIGA. It computes the sum of feature values in a given sample. The change in the sample is directly proportional to the sum of feature values.

5.3   Comparison Against Other Adversarial Models

Table 1 shows the computational complexity of our proposed approach in comparison with existing adversarial models. Adversarial model described in Sun and Upadhyaya [17] requires the attacker to iterate over the samples twice: first to look for wolves and

Table 1 Comparison of computational complexity

Adversarial model	Computational complexity
Negi and Sharma [14]	$O(n^2)$
Negi et al. [15]	$O(n)$
Sun and Upadhyaya [17]	$O(n^2)$
Our work	$O(n)$

lambs and second to forge the master key. Therefore, the computational complexity is $O(n^2)$.

Negi and Sharma [14] iterates over the samples to generate samples after creating clusters using k-means algorithm. Thus, the computational complexity of the adversarial model is $O(n)$. Similarly, Adversarial model in Negi et al. [15] uses the k-mean++ algorithm to generate samples. The algorithm iterates over the input and length of the password to generate samples. Thus the computational complexity is $O(n^2)$. In our work, we iterate over the samples only once and hence the computational complexity of our proposed model is $O(n)$.

6 Conclusion

In this paper, we have developed an adversarial model for keystroke authentication in embedded devices. The proposed approach leveraged the modified FIGA [1] algorithm to add perturbations to the typing pattern of a user due to which the error rate increases. Evaluation of the algorithm showed an error rate of 60, 55, and 63% for Random Forest, XGBoost, and SVM, respectively. Further, we performed the analysis of the overhead and the crafted adversarial samples. Analysis showed that it requires an average of 3.18 s to generate samples per user and that the average change in generated samples across all users is 1 s. Since the time taken to generate samples is reasonable and the change in those samples is minimal, our proposed model is observed to be feasible and realistic. The computational complexity of our adversarial model is $O(n)$ which is comparable with existing adversarial models. An attacker can leverage our adversarial model to generate samples that can successfully attack the mechanism of keystroke authentication in embedded devices.

In our future work, we will be exploring the effect of using our adversarial model in other biometric-based authentication systems such as touchscreen swipes, mouse movements, etc. In addition, we propose to develop a robust defence mechanism against the adversarial model to build resilient keystroke authentication systems.

.

References

1. G. Gressel, N. Hegde, A. Sreekumar, M. Darling, Feature importance guided attack: a model agnostic adversarial attack (2021). https://arxiv.org/abs/2106.14815
2. Embedded system market by hardware (MPU, MCU, application-specific integrated circuits,DSP, FPGA, and memories), software (middleware, operating systems), system size, functionality, application, region—global forecast to 2025 (Mar 2020). https://www.marketsandmarkets.com/Market-Reports/embedded-system-market-98154672.html
3. E. Anthi, L. Williams, M. Rhode, P. Burnap, A. Wedgbury, Adversarial attacks on machine learning cybersecurity defences in industrial control systems. J. Inf. Secur. Appl. **58**, 102717 (2021). https://www.sciencedirect.com/science/article/pii/S2214212620308607
4. A. Kaur, K. Mustafa, A critical appraisal on password based authentication. Int. J. Comput. Net. Inf. Secur. **11**, 47–61 (2019)

5. B. Schneier, Two-factor authentication: too little, too late. Commun. ACM **48**(4), 136 (2005). https://doi.org/10.1145/1053291.1053327
6. A.K. Jain, K. Nandakumar, Biometric authentication: system security and user privacy. IEEE Comput. **45**(11), 87–92 (2012)
7. K. Nimmy, S. Sankaran, K. Achuthan, A novel multi-factor authentication protocol for smart home environments, in *Information Systems Security*. ed. by V. Ganapathy, T. Jaeger, R. Shyamasundar (Springer International Publishing, Cham, 2018), pp. 44–63
8. A. Foresi, R. Samavi, User authentication using keystroke dynamics via crowdsourcing, in *2019 17th International Conference on Privacy, Security and Trust (PST)*, pp. 1–3. ISSN: 2643-4202
9. Amazon mechanical turk. https://www.mturk.com/
10. S. Singh, A. Inamdar, A. Kore, A. Pawar, Analysis of algorithms for user authentication using keystroke dynamics, in *2020 International Conference on Communication and Signal Processing (ICCSP)*, pp. 0337–0341
11. K.S. Killourhy, R.A. Maxion, Comparing anomaly-detection algorithms for keystroke dynamics, in *2009 IEEE/IFIP International Conference on Dependable Systems Networks*, pp. 125–134 (2009)
12. A. Daribay, M.S. Obaidat, P.V., Krishna, Analysis of authentication system based on keystroke dynamics, in *2019 International Conference on Computer, Information and Telecommunication Systems (CITS)*, pp. 1–6
13. S. Ravindran, C. Gautam, A. Tiwari, Keystroke user recognition through extreme learning machine and evolving cluster method, in *2015 IEEE International Conference on Computational Intelligence and Computing Research (ICCIC)*, pp. 1–5
14. P. Negi, A. Sharma, *Adversarial machine learning against keystroke dynamics*. Paper presented at Stanford; CS229 (2017)
15. P. Negi, P. Sharma, V. S. Jain, B. Bahmani, K-means++ versus behavioral biometrics: one loop to rule them all, in *Proceedings 2018 Network and Distributed System Security Symposium* (Internet Society, San Diego, CA, 2018)
16. H. Khan, U. Hengartner, D. Vogel, Mimicry attacks on smartphone keystroke authentication. ACM Trans. Priv. Secur. **23**(1), 2:1–2:34 (2020)
17. Y. Sun, S. Upadhyaya, Synthetic forgery attack against continuous keystroke authentication systems, in *2018 27th International Conference on Computer Communication and Networks (ICCCN)* (IEEE, Hangzhou, July 2018), pp. 1–7
18. F. Pedregosa, G. Varoquaux, A. Gramfort, V. Michel, B. Thirion, O. Grisel, M. Blondel, P. Prettenhofer, R. Weiss, V. Dubourg, J. Vanderplas, A. Passos, D. Cournapeau, M. Brucher, M. Perrot, E. Duchesnay, Scikit-learn: machine learning in Python. J. Mach. Learn. Res. **12**, 2825–2830 (2011)

A Multi-phase LC-Ring-Based Voltage Controlled Oscillator

Sounak Das and Subhajit Sen

Abstract This paper presents the design of a Voltage Controlled Oscillator (VCO) that achieves low phase noise as well as a wide tuning range that has been used to design the receive PLL clock of an OC-1920 SERDES system. The two popular VCO topologies, i.e., the LC-VCO and ring-VCO (ring-oscillator-based VCO) are compared with respect to tuning range as well as phase noise with LC-VCO showing a lower phase noise at the expense of poor tuning range as compared to a ring-VCO. In this paper a novel VCO circuit is proposed in which the resistive load of a ring-VCO is modified by adding an inductor in series with the load resistance. It is shown that this improves the effective Q of the VCO and therefore its phase noise. At the same time the tuning structure of the VCO is modified to achieve a higher tuning range. The VCO circuit achieves a tuning range of nearly 800 MHz with a phase noise of around −94 dBc/Hz at 1 MHz from the carrier. The multi-phase VCO is then embedded within an NRZ CDR (Clock-and-Data Recovery) to demonstrate the operation of an OC-1920 SERDES designed for a baud rate of 70 Gbps.

Keywords OC-1920 · Q · LC-VCO · VCO · CDR · SERDES · Phase Noise · LC tank · Ring oscillator

1 Introduction

A voltage controlled oscillator (VCO) is one of the main components in clock recovery systems. It produces a periodic output waveform which is fed back to phase detectors and then, depending upon the input, correction pulses are produced. These correction pulses are then used to lock the VCO frequency to the input data rate. The VCO implemented has been tested in the Clock Data Recovery (CDR) circuit of a

S. Das (✉)
A&MS Design Engineer, Synopsys India, Bangalore, India
e-mail: sounak.das@iiitb.org

S. Sen
IIIT, Bangalore, Bangalore, India
e-mail: subhajit_sen@iiitb.ac.in

high speed Serializer-deserializer (SERDES) system with baud rates in the order of 70 to 90 Gbps. Among many different implementations of a CDR, the phase locked loop (PLL) based CDR is chosen for its simplicity and widespread usage. To reduce the speed requirements of the digital blocks, a quarter-rate CDR architecture was used which requires a quarter-rate, four-phase VCO. The VCO is one of the main sources of jitter in a PLL-based system, where jitter is a direct function of phase noise exhibited by the VCO. LC tank-based VCOs exhibit very low phase noise if a coil with high quality factor (Q) can be implemented. However, their tuning range is low. On the other hand, ring-oscillator-based VCOs exhibit poorer phase noise but have higher frequency tuning ranges. Hence, there is a trade-off between phase noise and tuning range in VCOs. Keeping this in mind, a modified LC-ring-based oscillator has been conceived and implemented in this paper that exhibits a substantially better phase noise performance as compared to ring-based VCO's while providing a better frequency tuning range as compared to the LC tank-based VCOs.

This paper is arranged as follows. Section 2 briefly deals with the different types of VCO architectures, Sect. 3 describes the proposed VCO. All the architectures discussed are implemented using the 45 nm GPDK in Cadence™ and the results are based on schematic simulations. Section 4 deals with the quarter-rate CDR circuit, in which the proposed VCO is used for clock recovery. Results and comparisons are presented in Sect. 5 and the paper is concluded afterward.

2 Voltage Controlled Oscillator

2.1 An Eight-Phase Ring Oscillator

A ring oscillator is a chain of delay cells connected in a positive feedback configuration [1] as shown in Fig. 1.

Considering the delay of each cell being T_D, and there are N such cells, the frequency of the overall ring oscillator is

$$f_{osc} = \frac{1}{2NT_D} \tag{1}$$

There are many different implementations of the individual delay cells for the voltage controlled operation [1], most of which deal with actuating the load resistance of individual stages, which changes the overall delay and hence, the frequency

Fig. 1 Eight-phase Ring Oscillator

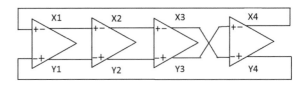

Fig. 2 Delay cell for a simple Ring Oscillator

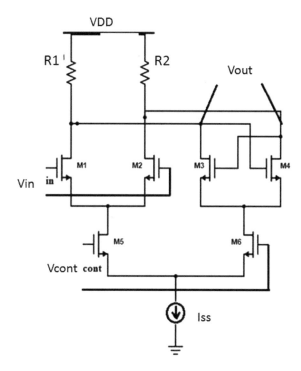

changes. Figure 2 shows one such implementation, which uses a negative resistance structure in parallel to the actual load of the delay cell. The negative resistance can be changed by changing the current through the cross-coupled pair which in turn changes the trans-conductance of the cross-coupled pair transistors.

The overall load resistance for a single delay cell is given as

$$R_P = \frac{R_{1,2}}{1 - g_{m3,4}R_{1,2}} \tag{2}$$

The phase noise of a typical ring-VCO depicted in Fig. 1 using the delay cell in Fig. 2 exhibits the effect of a very wide noise bandwidth, which is due to the low pass nature of the delay cells. The noise voltage injected by the load resistance, the cross-coupled transistors, and the driver transistors is shaped by a very wide noise transfer curve, which is low pass in nature due to the absence of any LC tank type structure in the circuit.

The tuning range of such an oscillator is very wide. Figure 3 depicts the large tuning range of a ring oscillator implemented using 45 nm GPDK technology from Cadence™.

Fig. 3 Tuning range of the Ring Oscillator implemented in 45 nm GPDK

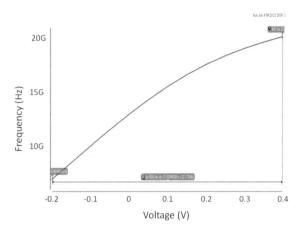

2.2 An Eight-Phase Ring Oscillator

The LC oscillator is a good choice when there is a requirement for low phase noise since it uses a high-Q LC tank. The noise injected by the series resistance of the inductor and the other transistor noise sources, are shaped by a band-pass noise-shaping function and hence, exhibit an improved phase noise as compared to the ring oscillator. The frequency of oscillations in case of an LC oscillator is dictated primarily by the LC tank resonance frequency and can be made to change in accordance to a control voltage using varactors. Four individual LC oscillator stages are usually coupled for implementing an eight-phase VCO, each of which is implemented as depicted in Fig. 4 [1].

The phase noise of an LC oscillator can be derived using the linear approach and it can be shown that the various noise sources are shaped by the following transfer function at the output of the oscillator [1].

$$\left| \frac{Y}{X}(j(\omega_0 + \Delta\omega)) \right|^2 = \frac{1}{4Q^2}\left(\frac{\omega_0}{\Delta\omega}\right)^2 \tag{3}$$

where X and Y are inputs and output applied to the open-loop oscillator model, ω_0 is the frequency of oscillation, $\Delta\omega$ is the frequency deviation and Q is the quality factor of the oscillator. In the case of N such stages, we can consider each of the stages acting as a band-pass filter with the band corner frequencies overlapping one another and thus exhibiting a phase noise, lower by a factor of N^2, as compared to the single LC oscillator stage. Hence, the multi-phase LC oscillator is superior to the Ring Oscillator, in terms of the phase noise because of the aforementioned chain of band-pass filtering facilitating a very high degree of noise filtering.

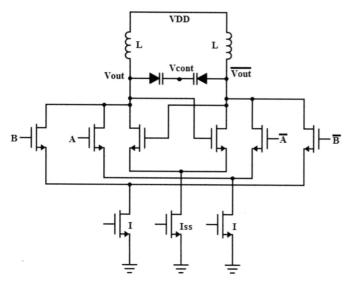

Fig. 4 LC Oscillator cell with coupling

3 Proposed VCO

For use in the CDR for high speed SERDES, there is a need to arrive at a trade-off between phase noise and tuning range. We do not require a high range of frequencies as we require the range of voltages for which the oscillator should start oscillating. In other words, a high voltage range for frequency tuning along with a moderate VCO gain is required for proper noise filtering inside a PLL-based system. In [2] a modified feed-forward interpolator (FFI) cell has been proposed for use in a quarter-rate receiver chipset. The proposed VCO architecture is a modification on [2], which uses a BiCMOS technology, since the use of BJTs is eliminated and the design can be fully implemented in any generic or state-of-the-art CMOS technology.

The phase noise of the proposed VCO is an improvement as compared to the ring-VCO due to the use of LC tank load. Tuning, in the FFI delay cell is achieved by controlling the current Itune using the differential control voltage which is also an advantage in our case as having differential control signals suppress the even harmonics which might have been there if a single-ended control line was used contributing to lesser systematic PLL phase noise generation.

The disadvantages in the FFI architecture come from the fact that the phase noise, for a given voltage swing is contributed by a greater number of active noise sources, some of which are used to couple the individual delay cells for a multi-phase operation, hence are inevitable. Also, as the control signals change from one extreme to the other, the operation of the VCO changes as follows; If the differential control signal is too low, one of the coupling transistors turns off and the VCO behaves as a pair of quadrature VCOs with no phase synchronization between the two.

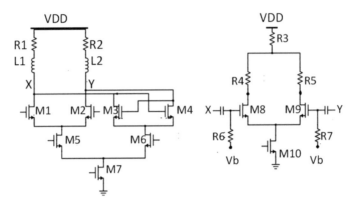

Fig. 5 Modified LC-Ring-VCO (left) and level-shifted output driver (right)

With the intention to alleviate the above issues, modifications were done to the delay cell of the Ring-VCO which is presented in Fig. 5. The delay cells are coupled in a ring configuration, the load is changed to a LC tank structure as opposed to a resistance and the tuning structure is simplified so that it uses lesser number of transistors and hence, fewer noise sources. The frequency is controlled using a differential control signal which changes the current passing through the cross-coupled transistor pair.

Considering the currents through the cross-coupled pair to be I_{SS} and the current through the inverting pair to be I. We can show that the phase difference between these two currents is exactly 45^0 [2]. The overall current passing through the active part of the delay cell, if considered to be I_T shows a phase shift of θ from I_{SS}. Then the following equation holds true.

$$\theta = tan^{-1} \frac{|I|\cos(\pi/4)}{|I_{SS}| + |I|\sin(\pi/4)} \tag{4}$$

V_{out} must align with the current through the cross-coupled pair, in that case, the phase shift introduced by the LC tank should cancel out the phase shift introduced in (4). Considering ω_{osc}, to be the oscillation frequency, according to [4], we can come up with the following equation.

$$\omega_{osc} = \frac{1}{\sqrt{LC}} \left[\frac{tan\theta + \sqrt{tan^2\theta + 4Q^2}}{2Q} \right], \tag{5}$$

where θ depends upon the currents through the inverting pair and the cross-coupled pair and hence, can be controlled using the differential control voltage.

According to [5], "open loop" Q is defined as

$$Q = \frac{\omega_0}{2} \left| \frac{d\varphi}{d\omega} \right| \tag{6}$$

From the Barkhausen criteria [3], considering an oscillating system to be a linear feedback system, the oscillation is sustained if the overall phase around the loop is 360°. Note that, only the effect of random noise introduced at the zero crossings is considered under the assumption that the VCO oscillation is hard-limited. If a noise is injected at the zero crossings, it violates the Barkhausen criteria and hence, the loop acts to restore it and the larger the slope of the $\varphi(j\omega)$, the quicker, this restoration happens and hence lesser the phase noise. Modifying the simple resistive load of a ring oscillator to LC tank structure increases this overall quality factor and hence, exhibits very low phase noise, coupled with the fact that there are three more such stages in cascade, the improvement in phase noise is phenomenal, while keeping a moderate frequency tuning range.

3.1 Proposed VCO Functionality and Results

The modified LC-Ring VCO is implemented using 45 nm GPDK technology available from Cadence. The inductor has been modeled as a series combination of a lossless inductor and a resistor, the value of which was obtained to be 100 mΩ per square This number is taken from the PDK Design Rule Manual (DRM) for the highest metal layer that the PDK can support. Inductors are generally implemented in the MTOP/MTOP-1 depending on the Redistribution Layer (RDL) used by the Digital Implementation (DI), so the resistivity of that layer was considered while modeling the parasitic resistance of the load inductor.

Figure 6 shows the transient waveform of the proposed VCO for a given control

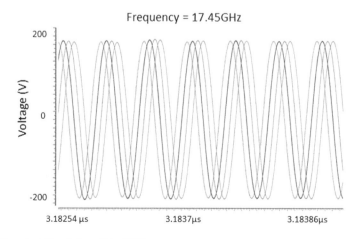

Fig. 6 Transient Simulation of the VCO

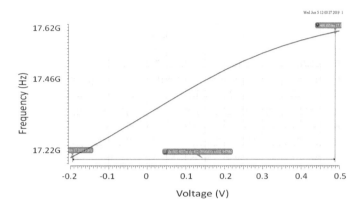

Fig. 7 Transfer characteristic of the proposed VCO

voltage. Figure 7 shows the transfer characteristics of the proposed VCO. The proposed VCO shows a frequency tuning range of more than 600 MHz when the differential control voltage is swept from −500 mV to 500 mV. A phase noise of −94 dBc/Hz is achieved at a spot frequency of 1 MHz from the carrier at 17.5 GHz output frequency. Figure 8 shows the phase noise plot of the VCO at an output frequency of about 17.5 GHz. Figure 9 depicts the comparative phase noise performance of the proposed VCO against a conventional LC and a conventional ring-VCO. The improvement in phase noise of the proposed (modified ring) VCO as compared with the ring-VCO is due to the use of an improved tuning structure as well as lesser number of active elements contributing lesser noise.

Fig. 8 Phase Noise of the proposed VCO

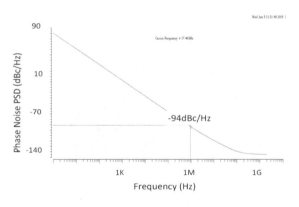

Fig. 9 Phase noise comparison of the three VCO architectures mentioned above

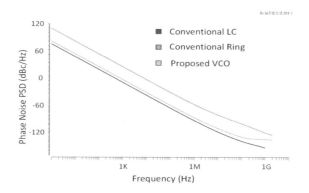

4 Quarter-Rate CDR

The proposed VCO is used in a Clock Data Recovery circuit designed for use in a high speed SERDES receiver block that achieves a baud rate of 70 GHz. A quarter-rate architecture reduces the speed requirement of the digital blocks. A bang-bang, PLL-based architecture is employed to minimize error in achieving phase lock due to the comparable digital block delays to that of the baud rate.

The phase detector of the CDR is implemented as presented in [4]. There are four pairs of UP/DWN signals from the phase detector which are used to provide phase correction for every fourth edge detected. The charge pump is a differential, negative-impedance amplifier (NIA) based implementation as described in [5] and [6]. The digital blocks are implemented using Current Mode Logic, as described in [1]. CML gates are fully differential and many times faster than the CMOS counterparts but exhibit a very small voltage swing. The design is intended for the OC-1920 band of optical communications standard. In the present implementation, a 70 Gbps system has been implemented using GPDK 45 nm CMOS technology. Figure 10 illustrates error-free performance of the CDR with an applied input data sequence. The applied input sequence of 1, 0, 0, 1, 1, 1, 1, 1, 1, 0, … gives the correct de-multiplexed output data of 1, 0, 0, 1, 1, 1, 1, 1, 1, 0, … and so on, as depicted in the waveform below.

5 Conclusions

In this paper a VCO is proposed which uses a Q enhancement load-based delay cell and a modified tuning structure to achieve acceptable phase noise and tuning range for applications in OC-1920 band of optical communications. The VCO has a tuning range of 600 MHz while simultaneously achieving a low phase noise of −94 dBc/Hz The proposed VCO is implemented and used in a OC-1920 rate CDR for high speed NRZ data with a baud rate of 70 Gbps. There is potential for the VCO to be used at significantly higher baud rates of up to 90 Gbps or above.

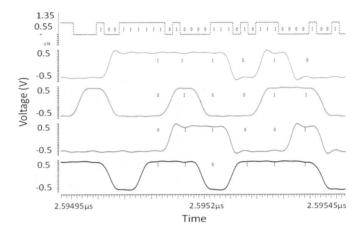

Fig. 10 Applied input data (top) and De-multiplexed CDR output data

In conclusion the proposed VCO has a better noise performance and poorer tuning range as compared to the ring oscillator implemented in the same PDK, while it has a better tuning range and poorer noise performance than the LC-VCO, implemented in the same PDK. The tuning range of the proposed VCO is still quite low, less than 1 MHz for a center frequency of 17.45 MHz (standalone). This can be improved by using digitally switched capacitor banks to coarsely change the center frequency and have the differential control signal to fine tune it to the required value.

The apple-to-apple comparison for power and jitter numbers w.r.t. the benchmark implementation is not present due to the lack of enough simulation and design data at this moment, which calls for a scope to further characterize the VCO at a standalone and PLL level as well.

With PVT variation data and schematic freeze, there is a scope to still better the performance by having a layout back-annotated schematic, which requires physical layout to be done. This is also not available at the moment.

Acknowledgements We are grateful to Prof. Chetan D. Parikh (IIIT Bangalore) and Prof. Madhav Rao (IIIT Bangalore) for their support and encouragement.

References

1. B. Razavi, *RF Microelectronics* (Pearson Education, 2012)
2. Y. Luo, *A high speed serializer/deserializer design*, Doctoral Dissertations (2010). https://sch olars.unh.edu/dissertation/536.
3. Z.B. Balaz, M. Minarik, V. Kudjak, V. Stofanik, Barkhausen criterion and another necessary condition for steady state oscillations existence, in *23rd International Conference Radioelektronika (RADIOELEKTRONIKA)*, Pardubice, pp. 151–155 (2013)

4. J. Lee, B. Razavi, A 40-gb/s clock and data recovery circuit in 0.18-μm CMOS technology. IEEE J. Solid-State Circuits **38**, 2181
5. D. Friedman, M. Meghelli, B. Parker, H. Ainspan, M. Soyuer, Sub-picosecond jitter SiGe BiCMOS transmit and receive PLLs for 12.5 Gbaud serial data communication, in *2000 Symposium on VLSI Circuits* (Digest of Technical Papers, Honolulu, USA, 2000), pp. 132–135
6. T.H. Lee, J.F. Bulzacchelli, A 155 MHz clock recovery delay- and phase-locked loop. ISSCC (1992)
7. T.H. Lee, *The Design of CMOS Radio-Frequency Integrated Circuits*
8. S. Kaeriyama et. al., A 40 Gb/s multi-data-rate CMOS transmitter and receiver chipset with SFI-5 interface for optical transmission systems. IEEE J. Solid-State Circuits (2010)
9. J.-C. Chien et. al., A 20-Gb/s 1: 2 demultiplexer with capacitive-splitting current-mode-logic latches. IEEE Trans. Microw. Theory Tech. (2007)
10. K. Watanabe et. al., A low-jitter 16:1 MUX and a high-sensitivity 1:16 DEMUX with integrated 39.8 to 43GHz VCO for OC-768 communication systems, in *2004 IEEE International Solid-State Circuits Conference, 2004. Digest of Technical Papers. ISSCC.* (2004)

Efficient Quantum Implementation of Majority-Based Full Adder Circuit Using Clifford+T-Group

Laxmidhar Biswal, Bappaditya Mondal, Anindita Chakraborty, and Hafizur Rahaman

Abstract The rapid progress in quantum technologies, as well as quantum algorithms, has paved the way for general-purpose, scalable quantum computers. The single most important challenge to that ambitious goal is the handling of noise. Literally, quantum states are fragile; highly sensitive toward noise which requires regular encoding and decoding of quantum information through quantum error-correcting code (QECC) so as to accomplish fault tolerance. Due to the efficient physical implementation of promising QECC on multiple quantum technologies, the Clifford+T group of Quantum operators is nowadays commonly used for the fault-tolerant Quantum circuit realizations. On the other hand, the majority-based data structure converts a ripple-carry adder into a carry-look ahead adder. In this work, we focus on the implementation of Majority-based 1-bit and n-bit Full Adder (FA) in the fault-tolerant quantum logic circuits. We have also calculated the performance parameters *viz. T-count*, ancillary cost, and *T-depth* that connected with the quantum circuit of full adder circuit in fault-tolerant logic.

Keywords Clifford+T · Full Adder · Majority · *T-count* · *T-depth*.

1 Introduction

Current CMOS technologies are at the verge of their limit due to power dissipation, atomic limit, and quantum tunneling, which pave the way for an alternative computational machine, such as quantum computer (QC) [1–3]. Basically, QC operates on the principle of quantum mechanics at the sub-atomic level and relies on qubit instead of classical bit [4]. In qubit, all classical basis states are incoherent superpo-

L. Biswal (✉) · A. Chakraborty · H. Rahaman
School of VLSI Technology, IIEST Shibpur, Shibpur, India
e-mail: hafizur@vlsi.iiests.ac.in

B. Mondal
Department of Computer Science & Engineering, Neotia Institute of Technology Management and Science, Kolkata, India

© The Author(s), under exclusive license to Springer Nature Singapore Pte Ltd. 2022   53
B. Mishra et al. (eds.), *Artificial Intelligence Driven Circuits and Systems*,
Lecture Notes in Electrical Engineering 811,
https://doi.org/10.1007/978-981-16-6940-8_5

sition called entanglement and allow parallel computation which makes the quantum computer more supremacy. And have the potential to solve the problems with exponential faster time which are classically intractable [5, 6]. Besides, quantum computer has many application in various disciplines *viz.* machine learning, searching, secure computing, cryptography, optimization, material science, condensed matter physics, quantum chemistry, and sampling [7]. However, the quantum hardware is highly prone to noise and causes loss of entanglement of qubit into a classical bit called decoherence [8] which makes the qubit more fragile. So, handling noise, and protecting entanglement is the foremost challenge in the scalability of the quantum computer which needs fault-tolerant quantum computation.

Fundamentally, fault-tolerance properties can be attained at the logical level of the quantum circuit by means of continuous encoding and decoding of physical qubit into logical qubit of quantum error-correcting code (QECC). And 2D-NN complaint surface code is the robust one with a threshold value of error rate per gate, i.e., 0.75% [9]. However, the rate of error per gate should be contained below the level of the threshold value of QECC so as to achieve scalability of the logic circuit synthesis, which could be attained by transversely and needs a set of universal operators [10]. In fact, Clifford+T is a well-known set of elementary primitive transversal operators which contains *viz.* non-Clifford T-gate, H-gate, $CNOT$-gate, Pauli's X, Y, Z-gate, S-gate and provides universality in order to design fault-tolerant quantum logic circuit [11–15]. Elements of Clifford+T-group and its properties are listed in Tables 1 and 2.

In another way, it can be stated that surface code along with a Clifford+T-group provides a default platform for designing a fault-tolerant quantum logic circuit which is the basis of a quantum algorithm. On the other hand, T-gate has high distillation cost along with high latency for which the fault-tolerant circuit with an optimized T-*depth* and T-*count* is more desirable.

Table 1 Primitive quantum operators and its properties

Name of primitive quantum operator	Schematic diagram	Properties	Transformation matrix	Name of primitive quantum operator	Schematic diagram	Properties	Transformation matrix
NOT (X)	⊕	$X\|0\rangle = \|1\rangle$ $X\|1\rangle = \|0\rangle$	$\begin{bmatrix} 0 & 1 \\ 1 & 0 \end{bmatrix}$	CNOT(CN)		$\|00\rangle \rightarrow \|00\rangle$ $\|01\rangle \rightarrow \|01\rangle$ $\|10\rangle \rightarrow \|11\rangle$ $\|11\rangle \rightarrow \|10\rangle$	$\begin{bmatrix} 1 & 0 & 0 & 0 \\ 0 & 1 & 0 & 0 \\ 0 & 0 & 0 & 1 \\ 0 & 0 & 1 & 0 \end{bmatrix}$
Z gate	Z	$Z\|0\rangle = \|0\rangle$ $Z\|1\rangle = -\|1\rangle$	$\begin{bmatrix} 0 & 0 \\ 0 & -1 \end{bmatrix}$	Hadamard(H)	H	$H\|0\rangle = \frac{1}{\sqrt{2}}(\|0\rangle + \|1\rangle)$ $H\|1\rangle = \frac{1}{\sqrt{2}}(\|0\rangle - \|1\rangle)$	$\frac{1}{\sqrt{2}}\begin{bmatrix} 1 & 1 \\ 1 & -1 \end{bmatrix}$
T gate	T	$T\|0\rangle = \|0\rangle$ $T\|1\rangle = e^{\frac{i\pi}{4}}\|1\rangle$	$\begin{bmatrix} 1 & 0 \\ 0 & e^{\frac{i\pi}{4}} \end{bmatrix}$	$T^\dagger$ gate	$T^\dagger$	$T^\dagger\|0\rangle = \|0\rangle$ $T^\dagger\|1\rangle = e^{\frac{-i\pi}{4}}\|1\rangle$	$\begin{bmatrix} 1 & 0 \\ 0 & e^{\frac{-i\pi}{4}} \end{bmatrix}$
S gate	S	$S\|0\rangle = \|0\rangle$ $S\|1\rangle = e^{\frac{i\pi}{2}}\|1\rangle$	$\begin{bmatrix} 1 & 0 \\ 0 & i \end{bmatrix}$	$S^\dagger$ gate	$S^\dagger$	$S^\dagger\|0\rangle = \|0\rangle$ $S^\dagger\|1\rangle = e^{\frac{-i\pi}{2}}\|1\rangle$	$\begin{bmatrix} 1 & 0 \\ 0 & e^{\frac{-i\pi}{2}} \end{bmatrix}$

Table 2 Clifford+T-based representation of NCV-based unitary gate

Clifford+T-based primitive quantum gate	Switching properties	Name of the NCV-based unitary quantum gate	Equivalent Clifford+T-based quantum circuit

Literally, a quantum computer works on quantum logic, which is a Boolean view of quantum mechanics. So, suitable intermediary data structures are under quest in the design automation industry for mapping from Boolean logic into quantum logic. In this regard, BDD, AIG, MIG, and XMG are well-known data structures through which input Boolean functions are processed for the synthesis of quantum logic [16]. And Majority-Inverter-Graph data structure, i.e., MIG is a potential data structure. In this conjecture, we would like to implement Full adder (FA) circuit over the majority function in the quantum domain using Clifford+T-group. Moreover, **FA** is an important constituent element of the Arithmetic Logic Unit (ALU). Besides, the majority-based full adder is more impressive than any other approach for handling large variable function in minimum steps. There exists scarcely any fault-tolerant realization of majority-based Full adder circuit in quantum logic yet albeit realization using QCA. However, there exist majority-based FA over reversible logic [17], in quantum logic [16].

The rest of the work is organized as follows: Preliminaries on the reversible and quantum logic circuits are given in Sect. 2. Our proposed approach is discussed in Sect. 3. And followed by the analysis of experimental results are in Sect. 4. At the end, the work is concluded in Sect. 5.

2 Preliminaries

Herein, the fundamentals about reversible and quantum logic are introduced for the better apprehension of the work. We have also discussed the associated cost parameters to find the performance of the fault-tolerant quantum logic circuit.

Definition 1 *(Qubit)* It represents a multi-state of information and the quantum view of the classical bit in quantum mechanics. Mathematically, a qubit is represented by a linear coherent superposition of basis states through complex probability in such a way that

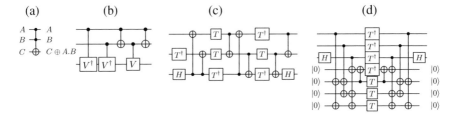

Fig. 1 **a** Toffoli-gate. **b** NCV-based Toffoli-gate. **c** Clifford+T-based Toffoli-gate [13]. **d** Unit T-depth based Toffoli-gate [19]

$$|\psi\rangle = \alpha|0\rangle + \beta|1\rangle \tag{1}$$

where $\alpha, \beta \in C$ and $|\alpha|^2 + |\beta|^2 = 1$.

Definition 2 *(Quantum gate)* It acts as an operator that is used to form the quantum circuits. The transfer matrix of quantum operators is unitary and Hermitian in nature. Due to the limitation of current quantum technologies, all elementary quantum operators of an universal gate set are either 4×4 or 2×2 transfer matrices.

Definition 3 *(NCV gate library)* It is a set of elementary primitive quantum gates, and universal toward the realization of any reversible gates which contains unitary operators *viz.* $CV, CV^\dagger, NOT$, and $CNOT$-gates. This gate library was introduced by Barenco et al. [18].

Definition 4 *(T-count)* Total number of T-gates/T-Matrices are available in the quantum circuit.

Definition 5 *(T-depth)* Total number of T-*cycle* needed to execute all the T-gates.

Definition 6 *(Toffoli-gate (TG))* It is a universal reversible gate that toggles the target bit when both the control input at high, i.e., logic '1'. On decomposition of Toffoli-gate into NCV-base elementary quantum operators, it has a quantum cost of 5. The schematic diagram of Toffoli-gate in reversible logic, and in quantum logic (NCV, Clifford+T-group) is depicted in Fig. 1.

2.1 Majority Function

The majority function maps n input Boolean variables into single output, is denoted as $\langle x_1 x_2, \ldots, x_n \rangle$, and evaluates to be true as '1' if and only if at least $\frac{n}{2}$ numbers of variables out of n-variables are true for even n and at least $\frac{n+1}{2}$ numbers of variables are true in case of odd n otherwise; the majority function shows false values as '0' [20]. Basically, the majority function has self-duality property and can be

Fig. 2 Schematic
representation 3-input
Majority function

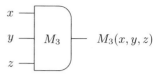

expected in the form of a conjunctive and disjunctive operator. Figure 2 presents a
schematic of the majority function ($\langle xyz \rangle$) for three input variables *viz.* x, y and z.
And Mathematically, it can be expressed as

$$\langle xyz \rangle = M(x, y, z) = xy + yz + zx \tag{2}$$
$$= x \wedge y \vee y \wedge z \vee z \wedge x \tag{3}$$
$$= \begin{cases} x \wedge y \; z = 0 \\ x \vee y \; z = 1 \end{cases} \tag{4}$$

2.2 Full Adder Circuit

It is a combinational logic circuit which adds three input bits and provides two output
bit as '*Sum*' bit and '*Carry*' bit. For example, the three inputs A, B and C_{in}, the output
Sum (S) and **Carry(C_{out})** can be described as

$$S = A \oplus B \oplus C_{in}$$
$$C_{out} = AB + BC_{in} + C_{in}A \tag{5}$$

3 Proposed Methodology

In this section, at first, we implement a quantum circuit for a majority-based 1-bit
Full Adder in fault-tolerant logic and then generalize our design approach for the
implementation of n-bit full adder.

3.1 Fault-Tolerant Quantum Implementation of
Majority-Based 1-Bit Full Adder

It can be implemented by mapping Boolean equation of FA circuit, i.e., Eq. 5 into
the Clifford+T-group through intermediary steps. Fundamentally, quantum compu-

tations are inherently reversible due to unitary which implies reversible logic is an intermediary logic in the aforesaid mapping. Though there exist multiple ways to map the Boolean logic into the reversible logic, we use a majority-based data structure to complete the task where the nodes are represented by the equivalent Toffoli-based reversible circuit.

Now, the outputs (S, C_{out}) of the FA (Eq. 5) are processed through the majority-based data structure and the resulting Mathematical expression is given by

$$
\begin{aligned}
S &= A \oplus B \oplus C_{in} \\
&= \overline{A}\overline{B}C_{in} + \overline{A}B\overline{C_{in}} + A\overline{B}\overline{C_{in}} + ABC_{in} \\
&= M_3(A, M_3(\overline{A}, B, C_{in}), \overline{M_3(A, B, C_{in})})
\end{aligned}
\tag{6}
$$

$$
\begin{aligned}
C_{out} &= AB + BC_{in} + C_{in}A \\
&= M_3(A, B, C_{in})
\end{aligned}
\tag{7}
$$

Figure 3 presents equivalent full adder circuit using 3-input majority-based block diagram. Fundamentally, if each small element of any Boolean logic is succeeded by a equivalent circuit in reversible logic then the resulting circuit becomes reversible circuit. So, all the 3-input majority blocks of Fig. 3 is to be replaced by its equivalent circuit in reversible logic. In this regard, we use the reversible circuit proposed by Chattopadhyay et al. [17, Fig. 3] which is an optimized one. Applying the replacement approach to Fig. 3 where each Mi_3 is succeeded by either Fig. 4a or b, the equivalent reversible circuit is presented in Fig. 5. And the resulting reversible circuit contains three Toffoli-gate and CNOT-gates which are to be decomposed into the primitive quantum operators of Clifford+T-group in order to realize an equivalent fault-tolerant quantum circuit.

For ease of decomposition approach, we use template matching schemes where each reversible Toffoli-gates are replaced by either Fig. 1c or d separately. It is important to note here that both Fig. 1c, d represent fault-tolerant architecture of Toffoli-gate and considered as Templates *viz.* T_1 and T_2, respectively.

Fig. 3 Majority-based full adder circuit using Eq. 6

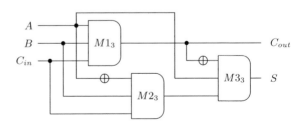

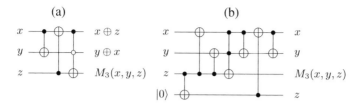

Fig. 4 **a** 3-input Majority function in reversible logic. **b** 3-input Majority function in reversible logic with ancillary line

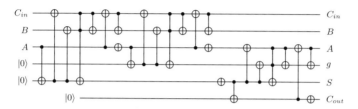

Fig. 5 Mapping of majority-based full adder into reversible logic

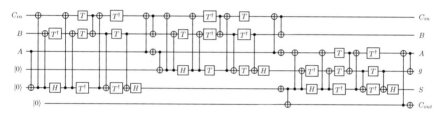

Fig. 6 Clifford+T-based quantum circuit realization of 1-bit full adder in fault-tolerant logic using template (T_1)

For template matching, we use the Algorithm 1 and traverse Fig. 5 from bottom-to-top, left-to-right approach. And replace each Toffoli-gate by its equivalent Clifford+T-based architecture that depicted in Fig. 1c, i.e., T_1. After due replacement, the out-coming quantum circuit is presented in Fig. 6. It has incurred 6 T-*depth* and 21 T-*count* as both $M1_3$ and $M2_3$ can execute independently and parallelly.

Again, a similar approach is followed by using a template (T_2) to realize quantum circuit in fault-tolerant logic for full adder circuit and the output circuit is given in Fig. 7. And the resulting circuit has obtained 21 T-*count*, and 2 T-*depth* in lieu of 6 T-*depth* with an additional 4 ancillary costs.

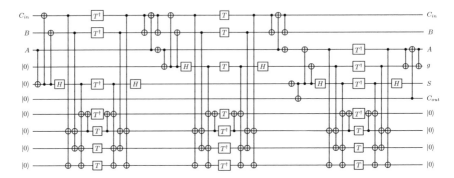

Fig. 7 Clifford+T-based quantum circuit realization of 1-bit full adder in fault-tolerant logic using template (T_2) with four additional ancillary costs

---

Algorithm 1: Reversible Circuit to Fault-tolerant Quantum Circuit Representation

---

    Input : Reversible Circuit(R_{ckt}),Template Library (T_{lib})
    Output: Clifford+T-based design
1 **begin**
2   T_{lib}=Include_ All$\{T_1,T_2\}$
3   **do**
4   initialization;
       Result: Initiate-scanning(R_{ckt},T_{lib})
5   **if** *(Result == 1)* **then**
6   Initiate_replacement(R_{ckt},$T_{selected}$)
7   Upgrade(R_{ckt});
8   **end**
9   **while** $\big($*Replacements are Possible*$\big)$;
10 **end**

3.2   Fault-Tolerant Quantum Implementation of Majority-Based n-bit Full Adder

From preceding Sect. 3.1, it can be drawn that C_{out} needs one M_3 with node depth 1 whereas; S-bit needs 3 numbers of M_3 in node depth 2 to execute/process. Further, the M_3 node that is used in the C_{out}, is reused in the realization of S that leads to a total of 3 numbers M_3 as needed for the realization of 1-bit full adder. Similarly, it can be stated axiomatically that n-bit full adder needs $3n$ numbers majority nodes of M_3 to execute/process. Due to limited space and pages, we skip the fault-tolerant quantum circuit diagram of n-bit full adder. However, we analyze the performance parameters associated with the realization of an equivalent fault-tolerant quantum implementation of n-bit full adder.

Calculation of T-count for n-bit full adder(TC(n))

Here we calculate T-count for n-bit full adder template wise separately. It is important to note here that both the templates have equal T-count. Further, each n-bit full adder contains a cascade of n numbers of 1-bit full adder where each 1-bit full adder follows the C_{out} of each preceding 1-bit full adder. From Sect. 3.1, each 1-bit FA has incurred 21 T-count while mapping into Clifford+T-group. So, the minimum numbers of T-gate needed to realize quantum circuit for n-bit FA in fault-tolerant logic is given by

$$TC(n) = n * TC(1)$$
$$= 21n \qquad (8)$$

Calculation of T-depth for n-bit full adder(TD(n))

Each T_1-based 1-bit full adder needs 6 T-cycle to operate all T-gate, i.e., Figure 6. Again, n numbers of 1-bit full adder are cascaded in n-bit full adder. So, TD (n) [T_1] the total numbers of T-cycle needed to run all T-gates that presented in the T_1-based fault-tolerant quantum circuit of n-bit FA are given by

$$TD(n)[T_1] = n * TD(1)$$
$$= 6n \qquad (9)$$

Similarly, TD(n)[T_2] the total numbers of T-cycle needed to operate all T-gate of fault-tolerant quantum circuit of T_2-based n-bit FA is given by

$$TD(n)[T_2] = n * TD(1)$$
$$= 2n \qquad (10)$$

4 Experimental Results

The experimental results, i.e., the performance parameters *viz.* T-count, T-depth, and ancillary cost which are related with the design of fault-tolerant quantum circuit of 1-bit FA and n-bit FA, have been evaluated in the preceding section.

As there hardly exists any direct mapping technique for deriving fault-tolerant realization of majority-based full adder so we are unable to correlate our result with the present state-of-the-art approaches. However, we highlight few advantages and drawbacks of our proposed design.

Basically, ripple-carry adder has major drawbacks due to the delay of the carry bit. However, the *Carry* is much faster than the *Sum* which is a novelty of the majority-based full adder circuit. Besides, majority-based n-bit full adder needs a lesser steps in comparison to its counter part in BDD, AIG, and ESOP data structure.

In the design approach, we have used two different kinds of templates against each reversible Toffoli-gate. Though both the design unaltered T-count but, template T_2 reduced the T-depth which is a major challenge for high scalable quantum computation in presence of decoherence. As all the 1-bit full adders are cascaded, so four additional ancillary lines can be reused by all 1-bit full adders which leads to a potential reduction of T-depth by a factor of 3 with four ancilla cost only.

We have also ensured the intactness of no-cloning [21] property in our designs as we have followed matrix identity property in the decomposition of large unitary function into primitive elementary forms.

5 Conclusion

Quantum circuit for 1-bit and n-bit full adder in fault-tolerant logic has been implemented using Clifford+T library. The performance parameters *viz.* T-count, and T-depth are optimized while mapping of majority-based reversible FA into fault-tolerant architectures. Besides, upper limit value of the performance parameters is also calculated for n-bit full adder. Besides, T-depth optimized templates are realized and also used so as to improve the scalability of the synthesis.

References

1. G.E. Moore, Cramming more components onto integrated circuits, reprinted from electronics, volume 38, number 8, Apr. 19, 1965, p. 114 ff. IEEE Solid-State Circ. Soc. Newsl. **11**(3), 33–35 (2006)
2. R. Landauer. Irreversibility and heat generation in the computing process. IBM J. Res. Dev. **5**(3) (1961)
3. C.H. Bennett, Logical reversibility of computation. IBM J. Res. Dev. **17**(6) (1973)
4. R.P. Feynman, Quantum mechanical computers. Found. Phys. **16**(6) (1986)
5. P.W. Shor, Polynomial-time algorithms for prime factorization and discrete logarithms on a quantum computer. SIAM Rev. **41**(2), 303–332 (1999)
6. L.K. Grover, Quantum mechanics helps in searching for a needle in a haystack. Phys. Rev. Lett. **79**(2), 325–328 (1997)
7. Quantum computing's promise for the brave new world (2019). https://blog.usejournal.com/quantum-computings-promise-for-the-brave&-new-world-fa15b651cced. Accessed 26 May 2019
8. D.P. DiVincenzo, Ibm. The physical implementation of quantum computation **48** (2000). https://www.onlinelibrary.wiley.com
9. A.G. Fowler, M. Mariantoni, J.M. Martinis, A.N. Cleland, Surface codes: towards practical large-scale quantum computation. Phys. Rev. A **86** (2012)
10. A. Paetznick, B.W. Reichardt, Universal fault-tolerant quantum computation with only transversal gates and error correction. Phys. Rev. Lett. **111** (2013)

11. L. Biswal, C. Bandyopadhyay, H. Rahaman, Efficient implementation of fault-tolerant 4:1 quantum multiplexer (qmux) using clifford+t-group, in *2019 IEEE International Symposium on Smart Electronic Systems (iSES) (Formerly iNiS)*, pp. 69–74, Dec 2019
12. L. Biswal, A. Bhattacharjee, R. Das, G. Thirunavukarasu, H. Rahaman, Quantum domain design of cliffordbased bidirectional barrel shifter, in *VLSI Design and Test*, ed. by S. Rajaram, N.B. Balamurugan, D. Gracia Nirmala Rani, V. Singh (Springer, Singapore, 2019), pp. 606–618
13. L. Biswal, C. Bandyopadhyay, A. Chattopadhyay, R. Wille, R. Drechsler, H. Rahaman, Nearest-neighbor and fault-tolerant quantum circuit implementation, in *2016 IEEE 46th International Symposium on Multiple-Valued Logic (ISMVL)*, pp. 156–161, (2016)
14. L. Biswal, K. Mondal, A. Bhattacharjee, H. Rahaman, Fault-tolerant quantum implementation of priority encoder circuit using clifford+t-group, in *2020 International Symposium on Devices, Circuits and Systems (ISDCS)*, pp. 1–6 (2020)
15. L. Biswal, C. Bandyopadhyay, H. Rahaman, clifford+t-based fault-tolerant quantum implementation of code converter circuit, in *Soft Computing: Theories and Applications. Advances in Intelligent Systems and Computing*, ed. by M. Pant, T. Kumar Sharma, R. Arya, B. Sahana, H. Zolfagharinia, vol. 1154 (2020)
16. L. Amarú, P. Gaillardon, A. Chattopadhyay, G. De Micheli, A sound and complete axiomatization of majority-n logic. IEEE Trans. Comput. **65**(9), 2889–2895 (2016)
17. A. Chattopadhyay, L. Amarú, M. Soeken, P. Gaillardon, G. De Micheli, Notes on majority boolean algebra, in *2016 IEEE 46th International Symposium on Multiple-Valued Logic (ISMVL)*, pp. 50–55 (2016)
18. A. Barenco, C.H. Bennett, R. Cleve, D.P. DiVincenzo, N. Margolus, P. Shor, T. Sleator, J.A. Smolin, H. Weinfurter, Elementary gates for quantum computation. Phys. Rev. A **52**, 3457–3467 (1995)
19. P. Selinger, Quantum circuits of t-depth one. Phys. Rev. A **87** (2013)
20. M.H. Moaiyeri, R.F. Mirzaee, K. Navi, T. Nikoubin, New high-performance majority function based full adders, in *2009 14th International CSI Computer Conference*, pp. 100–104 (2009)
21. D. Gottesman, Stabilizer codes and quantum error correction(phd thesis) (1997). Accessed 06 Jan 2021

Energy Efficient SRAM Design Using FinFETs and Potential Alteration Topology Schemes

Samarth Agarwal⑩ and **Rajeevan Chandel**⑩

Abstract Massive scaling calls for alternate technologies to sustain the future of electronics. FinFET technology has successfully emerged as an alternative for the planar CMOS technology and outperforms the latter. In this paper, FinFETs are used for designing 6 T SRAM cell. A comparative analysis is carried out with the standard CMOS technology. SRAMs consume more than eighty percent of the chip area and thus its optimization is of prime importance to all the VLSI SoC designers. A comprehensive analysis has been carried out on the basis of performance indices namely, average power consumed, propagation delay, static noise margin, and power delay product (PDP). Through the experimental analysis and simulation, it has been observed that on an average the energy dissipation or PDP for write and read operations in the FinFET based SRAM cell is nearly 73% lesser than the conventional CMOS based SRAM cells. Also, two innovative mechanisms namely Upper Voltage Lowering Technique (UVLT) and Lower Voltage Lifting Technique (LVLT) have been hereby devised which are topology oriented and thereby have implementation ease and high effectiveness which is in turn comparable with any other synthesis level process available in the industry. The reduction in percentage of leakage currents observed is clearly considerable i.e. 50.21% in case of UVLT and 45.64% in the case of LVLT, thereby illustrating the overall improvements observed and the massive future possibilities that these designs exhibit.

Keywords SRAM · FinFET · LVLT · Metal Oxide Semiconductor · Power Delay Product · Predictive Technology Models · Static Noise Margin · Short channel effects · UVLT

S. Agarwal (✉) · R. Chandel
National Institute of Technology Hamirpur, Hamirpur, HP, India
e-mail: rchandel@nith.ac.in

© The Author(s), under exclusive license to Springer Nature Singapore Pte Ltd. 2022
B. Mishra et al. (eds.), *Artificial Intelligence Driven Circuits and Systems*,
Lecture Notes in Electrical Engineering 811,
https://doi.org/10.1007/978-981-16-6940-8_6

65

1 Introduction

The ease of the read and write operations and the fulfilment of various performance requirements efficiently make the Static Random Access Memory (SRAM) an indispensable part of the memory design circuitry in Very Large Scale Integration (VLSI) technology. But as the transistor size gets scaled down to nanometer regime so as to get higher storage capabilities, Si based transistors have started to show up a heap of problems such as gate losing control of the channel, inability to shrink oxide thickness, higher power consumption in the idle state due to heavy leakage, and various short channel effects crop up. Particularly, at sizes as low as 20 nm and below, complementary metal oxide semiconductor (CMOS) can't be scaled down further [1]. Thus, to make the future electronic devices more computationally efficient and smaller, alternative materials and design concepts need to be opted for.

Researchers have tapped ever increasing performance and density out of the transistors being used in the integrated circuits, all thanks to reducing the size of the planarMOSFETs around last thirty years. However, due to massive rise in leakage at subthreshold level at technology nodes below 20 nm, a major hindrance occurs in taking ahead the Moore's Law. Due to extremely narrow lengths of the channel in nanoelectronics, the drain potential affects the channel electro -statistics and the device's gate exercises no control in operation [2]. This reduces the efficiency of the gate in the OFF mode leading to opening of channel path totally and thereby results in the rise of off currents. One way to alleviate this trouble is to use high k-dielectric materials and thinner gate oxides so as to increase the channel gate capacitance. However, there are many limitations to this approach as well. An alternative to the planar MOSFETs is the multiple gate FETs which by virtue of their extra gate, show good screening of the drain potential and also have increased channel gate capacitance [3]. Thus MGFETs are anyday better than MOSFETs when compared on the basis of various ShortChannel performance parameters like Drain Induced Barrier Lowering (DIBL), subthreshold slope, threshold voltage rolloff, etc. [4]. These clearly indicate lesser degradation in the threshold voltage of the device and the off current, with continued scaling. MGFETs like Trigate FETs and FinFETs have developed like the most desired replacement of the age old planar MOSFETs because of their simple structure and fabrication ease [5].

Over the past 10 years, FinFETs and Trigate devices have been thoroughly explored. A massive number of studies and articles are published that display the progress in the short-channel properties and behaviour as compared to traditional bulkMOSFETs [6]. Numerous specialists have provided novel circuit styles that utilize various kinds of FinFETs [7]. Academicians have additionally investigated different symmetric and unbalanced FinFET designs and utilized them in FinFET hybrid architectures and plans of memory designs. Later designs for processors and networks-on-chip (NoCs) have additionally been investigated. Regardless of these headways in FinFET research, articles that offer a worldwide perspective on FinFETs from the gadget level to the highest engineering level are scant. This paper is pointed toward an enormous assortment of readers: gadget engineers, circuit architects, and hardware designers. Our aim is to offer a worldwide perspective on FinFET standards spreading over the memory design progression and the designs of ICs. The rest of the paper is structured as follows. Section 2 discusses the structural details and cell operations of 6 T SRAM cell. Section 3 highlights the properties and parameters of FinFETs along with its advantages and disadvantages. Section 4 details the leakage current reduction techniques and discusses the low power techniques namely Upper Voltage Lowering Technique (UVLT) and Lower Voltage Lifting Technique (LVLT) in detail. Section 5 reveals the performance indices taken up for carrying out the feasibility analysis of the SRAMs. Section 6 presents the simulation results, their discussion and a comparative analysis has been carried out. Finally, some eminent conclusions are drawn in Sect. 7.

2   Static Random Access Memory

Static Random Access Memory (SRAM), is created using flip-flops. As long as the power is applied, it is able to retain its data and that is the reason which helps avoiding refreshing the hold state. The 6 T SRAM has 6 transistors, two control lines, one DC voltage supply, two bit lines (BL and BL_bar) and one address line as shown in Fig. 1. The SRAM read operation occurs as follows. The memory is say, $Q = 1$ and thus $Q\_bar = 0$. The Word Line (WL) = 1 for read/write operation and 0 for the hold operation. Both the Precharge_Capacitors are at voltage level say V_{DD}. The output at BL and BL_bar are obtained. The $Q\_bar = 0$ and node voltage is V_{DD}, so the capacitor discharges. The BL_bar voltage decreases and this downfall is detected by the sense amplifier. Finally, the comparator gives output as 1 i.e. required output of the read operation. Similarly, for write operation the following steps take place. The state of $Q = 0$ and $Q\_bar = 1$. The WL = 1 and BL and BL_bar will be the input

Fig. 1 A 6 T SRAM cell [1, 8]

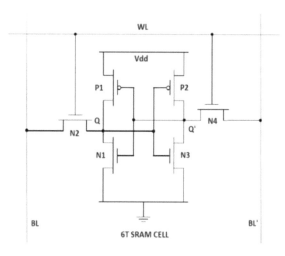

and will be controlled such that BL_bar is set to Gnd. BL_bar = Gnd and Q_bar = 1 so discharging occurs and voltage decreases. Therefore, as the voltage falls below the threshold voltage (V_{tn}) of N1, N1 turns off and thus P1 turns on setting up the output or Q as 1. Hence, the value of Q has successfully been changed from 0 to 1 i.e. the write operation is accomplished.

3   FinFETs

FinFET is a type of multi-gate MOSFET. It is called as FinFET on the grounds that the 3D design over the substrate looks like the fin arrangement in the fish. It is a 3D transistor and widely used in integration circuits recently instead of planar CMOS FETs. It is used more than other FETs because of its area of performance, lower leakage power, low voltage operation, intra die variability and lower retention voltages for SRAM. In a planar FET, the gate is placed above the channel and there is leakage current flowing from the source terminal to the drain terminal even when the gate is in off state. The structure of the FinFET consists of a drain, source and also a gate terminal to control the flow of current. The channel is in the shape of a thin vertical fin and gate is enveloped around it. This helps in better controlling of the channel and thus the electrical properties are better. FinFETs can have two or four fins in the same structure. Figure 2 represents a basic structure of FinFET.

The step-by-step manufacturing process is as follows. First step is substrate in the fabrication process i.e. a lightly doped p-type substrate is fabricated and a hard mask is fabricated over the substrate. Next step is fin etch i.e. by a highly anisotropic process, the fins are formed. Then the outside position over the fins' outside layers are formed to isolate the fins. Finally, it is planarized by chemical–mechanical polishing process. Next step is recess etching. By this process excess oxide is etched. Then the gate oxide is used to isolate the channel from the gate by thermal oxidization process. The gate oxide is placed on the fins. This is followed by the formation of the gate. Then, a heavily doped N+ polysilicon gate coating is formed and deposited over the fins. The current–voltage characteristics of FinFET show that the drain current increases when the drain source voltage is applied. Initially it increases linearly and after that it enters into the saturation region where the curve is almost constant. The various advantages of using FinFETs are: (i) low power consumption (ii) operates at low voltage (iii) operating speed is higher (iv) the leakage current (static) is decreased up to 80% (v) more compact. However, the disadvantages are (i) for building the FinFET it involves many additional steps so the fabrication cost is high (ii) controlling the fin depth is difficult. FinFETs are used in the various microprocessors, smartphones, and other faster and compact circuit chips.

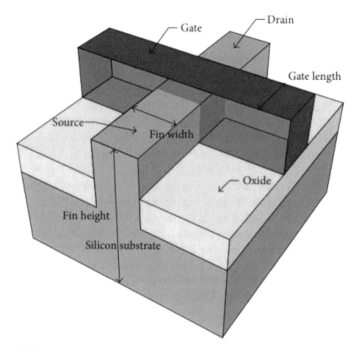

Fig. 2 A FinFET structure

4 Leakage Current Reduction Techniques

The leakage current occurs in an SRAM on account of the following mechanisms namely Subthreshold Leakage, GateDirect Tunneling Leakage, GateInduced Drain Leakage and Reverse Biased Junction Leakage. Curbing the ever-increasing power utilization is of prime importance to chip designers today and for that, all the components contributing should be taken into account. Some of the most considerable ones are reduction of capacitance, reduction of DC current and AC current, reduction of voltage of operation and leakage current in both standby and active state. This can be achieved using techniques viz Multithreshold CMOS (MTCMOS), Variable-threshold CMOS (VTCMOS), clock gating techniques, Multivoltage optimizations, Well Biasing method, DynamicVoltage and Frequency Scaling (DVFS), etc. Current leakage depletion techniques are a must whenever low power VLSI designing is considered. As a part of the research work, the following two schemes namely UVLT and LVLT have been thoroughly explored which are based on topological alteration. The circuits associated have been simulated using HSPICE and the associated transistor current measurements are recorded. These have then been compared with the current measurements obtained from a standard CMOS 6TSRAM cell and thereby the comparative analysis has been performed showing the percentage improvements elaborately.

4.1 Upper Voltage Lowering Technique

As illustrated in Fig. 3, the Upper Voltage Lowering technique (UVLT) is a great prospective leakage current reduction scheme. Full supply voltage is exerted in the active stage of memory cell. However, in the inactive stage, the applied voltage is reduced. With M4 being turned on, the voltage level at M4 and M2 is pulled down. The gate leakage current is reduced as the V_g of M1 falls. I_g through M2 also falls significantly. Voltage at source of M6 falls which leads to fall of Edge Direct Tunneling leakage. M5 leakage at gate remains unaltered. Thus, UVLT shows good results for reduction of leakage current at the gate terminal. However, it's not as effective rather is comparatively poor in case of subthreshold current leakage. As observed, in UVLT scheme, three additional transistors are put over the SRAM cell one of which is clocked. During the standby mode, all transistors in the cell show varied amounts of leakage currents based on the cell holdings. Say, if zero is kept inside, edge direct tunneling occurs through the N type transistors whereas for p type ones, the gate leakage is meagre. The subthreshold leakage occurs in the off state devices namely M3, M2 and M5. To curb the trouble caused by increased subthreshold leakage current, the LVLT strategy of cell designing has been taken into consideration.

Fig. 3 SRAM with UVLT [9, 10]

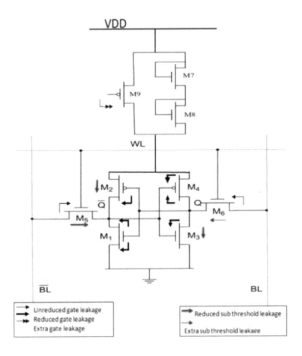

4.2 Lower Voltage Lifting Technique

If the ground potential is slightly increased in the standby stage, which in turn brings down the swing, and thereby leakage. As our major objective is to control power leakage, this idea can be exploited. Three transistors are hereby additionally added, one of which is clocked, used as a switch of control. The clock is posedged in the read–write stage and disabled in the standby stage to make way for an alternate current trail to flow. Owing to this, a decrease in the leakage of current is observed. Figure 4 illustrates the Lower Voltage Lifting Technique (LVLT) included SRAM circuit. During active stage, the switch gives zero volts at gnd and increases gnd level to Gnd$_{virtual}$ while in the inactive state. On carefully analyzing the impact of LVLT on gate leakage, it is observed that owing to decrease in V_{gs} and V_{gd} of M1 and V_{gd} of M2 lead to steep decrease in the gate leakage. On the contrary, there isn't any improvement for M6 and M5 pass transistors, respectively. Thus overall improvement in gate leakage isn't as effective as in the case of UVLT. On carefully studying the subthreshold leakage effect of LVLT, this idea emerges out as profitable in reducing the M5, M2 and M3 leakages substantially. This add-on circuitry should thus definitely be added as it gives a win–win situation for low power VLSI memory designers.

Fig. 4 SRAM with LVLT [9, 10]

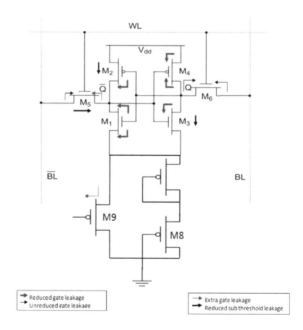

5 Performance Indices for the SRAM

The performance analysis has been experimented on the basis of four parameters namely delay, power dissipated, power delay product (PDP) and static noise margin (SNM). Delay basically gives us an idea about how fast the switching operations take place. The faster the operation, the better the transistor performance. In today's era, equipment speed is one of the major requirements [11]. Therefore, measuring the delay will help us analyse the efficiency of the product under test. This in-turn helps to determine the better technology. The power dissipation should be minimum for a good transistor device because more power dissipation leads to huge wastage of energy and energy is a very critical resource these days. The power delay product (PDP) is the multiplication of the delay and the power utilized. PDP represents the energy dissipation in a circuit. The lower it is, the better will be the robustness and the reliability of our memory design. The firmness and the writeability of the cell are enumerated by the following parameters namely, the hold margin, read margin and write margin which are resoluted by the SNM. It quantifies the amount of noise that can be applied at the input terminals of two cross combined inverters until a stable state is lost during the hold or read operating mode or an alternate state is produced that is stable during the write operation. The higher the SNM, the better will be the memory design [12]. In the present work, the efficiency of FinFETs based SRAM shall be further analysed and compared to that of the CMOS based SRAM cell.

6 Results and Discussion

The 6 T SRAM cell has been designed using the EDA simulator tool H-SPICE [13], and with the help of Predictive Technology Model 2012 (PTM-MG) files. The performance of various nanoscale technology nodes has been analysed for both FinFET and CMOS technology based SRAM cell. The comparison has been established on the basis of the various parameters and has been appropriately analysed. Additionally, the values of the leakage currents have been obtained from various transistors along with their prominent leakage considerations. The results have been obtained and tabulated for all the three SRAM configurations namely the conventional 6 T design, LVLT design and UVLT design, in Tables 1, 2, 3 and 4.

6.1 Comparative Analysis

From Table 1, it is evident that the propagation delay in case of FinFET based SRAM cell is much lower as compared to its CMOS based rival design. For instance, during the read operation for SRAM designed using 20 nm technology file, it is seen that in case of CMOS design, the delay is 142 ps, whereas for the same design when

Table 1 Propagation delay in picoseconds

SRAM built using	Operation mode	Technology node		
		32 nm	20 nm	16 nm
CMOS	Read delay (ps)	171	142	103
FinFET	Read delay (ps)	76.5	43	15.8
Improvement	Read delay (ps)	55%	70%	85%
CMOS	Write delay (ps)	183	162	122
FinFET	Write delay (ps)	128	41.2	35
Improvement	Write delay (ps)	30%	75%	72%

Table 2 Average power dissipation in nanowatts

SRAM built using	Operation mode	Technology node		
		32 nm	20 nm	16 nm
CMOS	Read power (nW)	75	189	233
FinFET	Read power (nW)	74	111	61
Improvement	Read power (nW)	1%	41%	74%
CMOS	Write power (nW)	75	209	297
FinFET	Write power (nW)	75	143	36
Improvement	Write power (nW)	0%	31%	87%

Table 3 Power delay product in attoJoule

SRAM built using	Operation mode	Technology node		
		32 nm	20 nm	16 nm
CMOS	Read PDP (aJ)	12.825	26.838	23.999
FinFET	Read PDP (aJ)	5.661	4.773	0.963
Improvement	Read PDP (aJ)	55%	82%	95%
CMOS	Write PDP (aJ)	13.725	33.858	36.234
FinFET	Write PDP (aJ)	9.600	5.891	1.260
Improvement	Write PDP (aJ)	30%	82%	96%

Table 4 Leakage current observed in various SRAM configurations

Transistor	Leakage current (in pA)	Standard SRAM cell	UVLT SRAM cell	LVLT SRAM cell
M1	Gate current	29.22	8.75	7.32
M2	Gate current	12.74	2.82	3.13
M3	Subthreshold current	0.93	0.67	0.21
M4	Subthreshold current	0.48	0.02	0.35
M5	Gate + subthreshold	12.82	11.92	11.15
M6	Gate current	18.37	12.94	18.37
Total current (pA)		74.56	37.12	40.53

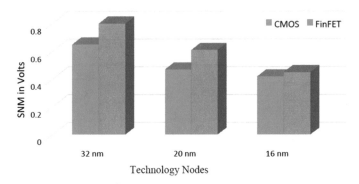

Fig. 5 SNM analysis of FinFET and CMOS based SRAMs

implemented using FinFET, the delay is 43 ps i.e. reduced by 70%. Similarly, for write mode and for other technology node sizes, a similar trend is observed. Also, from Table 2, the average power dissipated is also much less in case of the FinFET based cell as compared to its counterpart. For example, during the write operation for SRAM designed using 16 nm technology file, it is observed that in case of CMOS based design, the power dissipated is 297 nW, whereas it is 36 nW i.e. reduced by 87%, for the same design when implemented using FinFETs. A similar trend is observed for read mode and for other technology node sizes. Also, from Table 3, the values of PDP and their corresponding percentage reduction in values when the FinFET based SRAM design is compared with its CMOS based counterpart. Clearly, on an average, there is a massive reduction in PDP by 73.33%. Furthermore, these observations make our experiment worth it and it can be concluded that the paradigm shift from planar MOSFETs to 3D FETs is going to be extremely fruitful. Figure 5 shows that the values obtained in case of the ideal SNM computations for the FinFET based SRAM cell are higher than its planar MOS counterpart for all the three technology nodes considered. Therefore, it may be added to the inferences that the FinFET based design is definitely more robust and reliable. Along with that, Table 4 shows the values of the leakage currents that have been obtained for the various transistors along with the prominent leakage components. The results have been obtained and tabulated for all the three SRAM configurations namely the conventional 6 T, LVLT and UVLT designs. The overall percentage reduction observed after computations is clearly considerable i.e. 50.21% in case of UVLT and 45.64% in the case of LVLT, thereby illustrating the overall performance improvement and the massive future possibilities that these designs exhibit.

7 Conclusion

After all the simulations and comparative analysis, this paper highlights the performance of 6 T FinFET based SRAM cell at various nanoscale technologies. The comparison has been established with respect to the SNM and the PDP of the 6 T CMOS SRAM cell and the FinFET SRAM cell. The transient analysis has been performed for read, write and standby operations and the power and delay have been accordingly calculated. The results show a reduction in the Power Delay Product by an average of 82% in the write operation and by 87% for the read operation of FinFET based SRAM cell at 32 nm, 20 nm and 16 nm technology nodes, respectively as compared to its CMOS counterpart. Similarly, the results also show an improvement in the static noise margin by 23, 29 and 7%, respectively, which is quite considerable for all the VLSI design engineers. Thus, it can be conveniently proclaimed that the future of memory design lies in the core of the FinFET technology. This paper also enlightens us about the various works being carried out in the LowPower Circuit designing with pure highlight over SRAMs. Several techniques were taken into account and considered to bring down the current leakage without altering reliabilities and sustainabilities. Two innovative mechanisms namely Upper Voltage Lowering Technique and Lower Voltage Lifting Technique have been hereby devised. The UVLT and LVLT are topology oriented and thereby have implementation ease and high effectiveness which is in turn comparable with any other synthesis level process available in the industry. This research may serve as a great resource for both academicians and industrial designers so as to choose the right technology for their upcoming ventures.

References

1. M. Kang, S. Song, FinFET SRAM optimization with fin thickness and surface orientation. IEEE Trans. Electron Devices **57**(11), 2785–2793 (2010)
2. A. Bhoj, R. Joshi, Transport-analysis-based 3-D TCAD capacitance extraction for sub-32-nm SRAM structures. IEEE Electron Device Lett. **33**(2), 158–160 (2012)
3. A. Bhoj, R. Joshi, Efficient methodologies for 3-D TCAD modelling of emerging devices and circuits. IEEE Trans. CAD ICs Syst. **32**(1), 47–58 (2013)
4. N. Jha, Parasitics-aware design of symmetric and asymmetric gate-work function FinFET SRAMs. IEEE Trans. VLSI Sys. **22**(3), 548–561 (2014)
5. A. Carlson, Z. Guo, S. Balasubramanian, SRAM read/write margin enhancements using FinFETs. IEEE Trans. VLSI Syst. **18**(6), 887–900 (2015)
6. T. Park, H. Cho, J. Choe, Characteristics of CMOS SRAM cell using body-tied TG MOSFETs (Bulk FinFETs). IEEE Trans. Electron Dev. **53**(3), 481–487 (2016)
7. M. Samson, Static performance analysis of Low power SRAM. Int. J. Comput. Sci. Netw. Secur. **10**(5), 189–195 (2010)
8. E. Nowak, I. Aller, T. Ludwig, Turning silicon on its edge: double gate CMOS/FinFET technology. IEEE Circuits Devices Mag. **20**(1), 20–31 (2018)
9. Qin, H., Cao, Y., Markovic, D.: SRAM Leakage Suppression by Minimizing Standby Supply Voltage. Berkeley EECS Annual Research Symposium, UC Berkeley, USA (2004)
10. N. Gupta, R. Kumar, Design and Analysis of Low Power SRAM. Ph.D. thesis, TU, India (2012)
11. Kuhn, K. J.: CMOS scaling for the 22nm node and beyond: Device physics and technology, in *Proceedings of the International Symposium on VLSI-TSA*, pp. 1–2 (2011)
12. N. Sharma, R. Chandel, Variation tolerant and stability simulation of low power SRAM cell analysis using FGMOS. Int. J. Model. Simul. Sci. Comput. (2021)
13. HSPICE 2016: EDA tool by Synopsys

A Varactor-Based Impedance Tuner Circuit Operating at ISM Frequency Band Centered at 2.45 GHz

Chithra Liz Palson, Deepti Das Krishna, and Babita Roslind Jose

Abstract This paper presents a varactor-based impedance tuner operating at the industrial, scientific, and medical (ISM) frequency band centered at 2.45 GHz. The proposed impedance tuner is an L-section circuit with only three control voltages. It follows a linearized topology of anti-series varactor diode configuration with a center tap biasing. In addition to this, varactor stacks with $M \times M$ varactors in series $\times$ parallel form. Here, varactors are the tunable elements and an impedance inverter concept is used to implement variable inductance. Through a systematic optimization process, the tuner circuit is shown to have a good impedance tuner coverage with minimum number of control voltages.

Keywords Impedance tuner · RF adaptivity · Impedance matching network · Varactors

1 Introduction

Modern wireless transceiver technologies require adaptive impedance tuners that enable antenna mismatch correction [1–5], frequency reconfigurability or multi-band operation [6], tunable phase shift [7], and power amplifier efficiency improvement [6, 8]. In the current scenario, impedance tuners are used in portable systems like the handsets because they have electrically small antennas whose electromagnetic properties vary [5] according to the change in operating conditions, proximity with human beings or other electronic devices, interaction with other dielectrics, and so on. These variations can lead to antenna mismatches—increasing reflections and hence decreasing the efficiency of the whole communication system. Therefore, it

C. L. Palson (✉) · B. R. Jose
Division of Electronics Engineering, School of Engineering, Cochin University of Science and Technology, Kochi-39, India
e-mail: chithraliz@cusat.ac.in

D. D. Krishna
Department of Electronics, Cochin University of Science and Technology, Kochi-39, India

© The Author(s), under exclusive license to Springer Nature Singapore Pte Ltd. 2022   77
B. Mishra et al. (eds.), *Artificial Intelligence Driven Circuits and Systems*,
Lecture Notes in Electrical Engineering 811,
https://doi.org/10.1007/978-981-16-6940-8_7

would be highly desirable to implement a tunable impedance matching network that adaptively compensates the antenna impedance variations.

Toward this objective, several technologies have been used in the past few years. Among such, micro-electro-mechanical system (MEMS) devices [9, 10] are utilized in high-power applications in the frequency range from 5 to 110 GHz. Here, the tunability is achieved by using MEMS either as a switch or as a variable capacitor and such circuits have the advantage of linearity, lower power consumption, and low loss. However, there also come with several limitations such as mechanical wear and tear leading to a reduced lifetime, time delays in the order of microseconds, expensive when compared to other technologies and difficulties in packaging.

Compared to MEMS, hybrid technology [11], where surface mount device (SMD) components and distributed elements are mounted on a low-loss dielectric substrate, appeals as a more practical and economical approach. This is intended for applications in the frequency range 100 MHz–10 GHz. Here, tunability is achieved by using PIN diode switches or semiconductor varactors.

Moving on, Integrated Circuit (IC) technology [12, 13] has the advantage of reducing the dimensions of the components to the order of 1 mm. But significant losses can occur if not fabricated on low-loss substrates. Here, the tunability is obtained from components used in hybrid technology and switching transistors. Integration of MEMS-based components on standard silicon-based complementary metal-oxide-semiconductor (CMOS) technology has also been an area of research as this enables a high level of integration and lessens the manufacturing cost [14].

In this paper, we have proposed the design of an impedance tuner at the ISM frequency band centered at 2.45 GHz. We have used a hybrid approach with semiconductor varactors as the tunable elements. The design is simple and economical with minimum number of control voltages and maximum impedance tuner range.

2   Re-Configurable Impedance Tuner

In this section, the design of the proposed varactor-based impedance tuner is explained, followed by the description of the circuit topology containing the varactor devices with their equivalent model, the design constraints, and the optimization flow chart. Further, the simulation of the proposed circuit and results are explained.

2.1   Circuit Design

This paper aims to design a varactor-based impedance tuner network with a minimum number of control voltages at the ISM frequency band centered at 2.45 GHz. The various design requirements for proposed circuit include high Q factor, linearity, huge capacitance tuning range, high breakdown voltage, and small series resistance.

Fig. 1 SPICE model of
varactor [16]

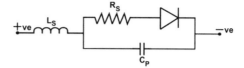

It is not possible to meet all these requirements simultaneously [15] and there is
always some trade-off between them.

Varactor Diode In the proposed design, we have used a silicon hyper abrupt junction
varactor diode, SKYWORKS SMV1233 [16] which is known for its low series resistance and high capacitance ratio. Figure 1 shows its SPICE model where the series
resistance R_S is 1.2 ohms, packaging capacitance C_P- 0.81pF, and series inductance
L_S- 0.7 nH. Theoretically, a varactor capacitance varies inversely with its applied
reverse voltage, and the relation between the capacitance of the diode $C_V(V_r)$ and
the applied reverse bias voltage V_r is given by

$$C_v(V_r) = \frac{C_{j0}}{(1 + \frac{V_r}{V_j})^\gamma} + C_p \qquad (1)$$

Here, the junction capacitance of the varactor at zero potential C_{j0} is 4.21 pF, the
built-in potential V_j is 11.87 V, and the grading coefficient γ is 6.43. The capacitance
of the diode SMV1233 varies from 5.08 pF to 0.84 pF as the voltage changes from
0 V to 15 V.

Circuit Development Figure 2 shows the proposed design. Here, C_s represents a
2×2 anti-series varactor pair stack, C_p represents an anti-series varactor pair, and L_p
represents a variable inductor formed by a transmission line loaded with a grounded
2×2 anti-series varactor pair stack. There are only three control voltages to bias the
varactor stacks. The biasing circuitry has a DC feed inductor L_F to block AC, anti-
parallel varactor pair to provide high zero-bias impedance, and DC block capacitors
C_{dc} to block DC.

Linearity There are different topologies reported to increase the linearity of varactor-
based networks. Of these available ones [17–19], an anti-series varactor pair, which
functions as a single tunable capacitor, is shown to improve the linearity. An anti-
parallel varactor pair configuration is connected at the center tap connection of the
anti-series varactor pair to provide an extremely high impedance for the dc biasing
supply. Thus, this anti-series varactor stack topology acts as a linear capacitor and
its capacitance can be tuned continuously by varying the dc bias voltage applied at
its center tap. This is incorporated in the proposed circuit.

Turn-on and Breakdown voltage limits In addition to the above-discussed constraints,
turn-on and break down of the varactor should be avoided in order to avoid high losses
and to ensure survival. The solution is to use varactor stacks with $M \times M$ varactors
in series $\times$ parallel configuration [8, 20]. This type of varactor stack functions as
a single variable capacitor element and divides the RF voltage swing across each

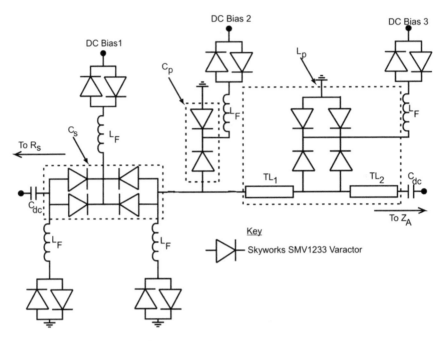

Fig. 2 Varactor-based adaptive impedance tuner

varactor junction by a factor of M. Hence, it enhances the equivalent breakdown voltage and the tuning factor of the varactor. In this paper, we have used a 2×2 varactor stack C_s as shown in Fig. 2.

Variable Inductor As we aim to realize a tunable impedance matching network, several tunable components such as variable capacitors and inductors are required. In the proposed design, varactors are used as the variable capacitors, however, practical variable inductors are not readily available. There are two methods to realize variable inductance: One is to connect a variable capacitor (varactor) in series or parallel with a fixed inductor, and the second to use the idea of impedance inverter using a varactor and a quarter-wave transmission line as shown in Fig. 3. The latter is used in the proposed circuit to obtain variable inductance, L_p as shown in Fig. 2. The circuit uses an FR_4 substrate with $\epsilon_R = 4.7$ of 1.25 mm height and double layer copper laminate.

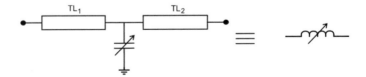

Fig. 3 Variable inductor

2.2 Design and Optimization

One of the important parameters of a tunable impedance network is its tuning range. The percentage area covered by the matched impedances on a Smith chart gives a clear idea about the tunable range of an impedance tuner [11, 21]. A design optimization process is followed for minimum return and insertion losses. As varactors are the tunable elements, there are also constraints based on the voltage across the varactor junction. First, the maximum voltage across the varactor junction must be less than its breakdown voltage, and second, the minimum voltage across the varactor junction should be less than the forward voltage. Hence the following are the design constraints for the proposed varactor-based tuner circuit.

- Return Loss < -10 dB
- Insertion Loss < 3 dB
- Maximum Voltage across the varactor junction $<$ Breakdown voltage, V_{bd}
- Minimum Voltage across the varactor junction $<$ Forward voltage, V_{fd}

An optimization process followed based on the above constraints is shown in Fig. 4. During the process, the source impedance is fixed as 50 ohms and a set of

Fig. 4 Flowchart for the optimization process

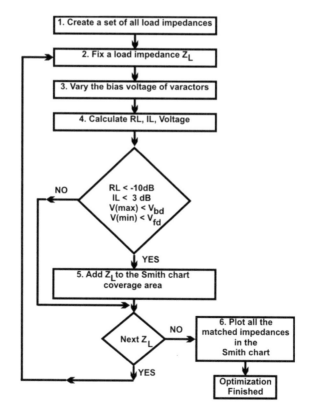

uniformly distributed load impedances on the Smith chart is chosen. For all the load impedances, the steps shown in Fig. 4 is followed. The ones which could be matched under the given constraints are plotted on the Smith chart and thus giving the visual representation of the tuner coverage.

2.3 Simulated Results

The proposed varactor-based matching network is simulated in Keysight Advanced Design System (ADS) along with the varactor SPICE model shown in Fig. 1. Here FR_4 substrate with $\epsilon_R = 4.7$ of 1.25 mm height and double layer copper laminate is used. As discussed, the source impedance of the network is fixed at 50 ohms and the load impedance is varied. After the optimization process as discussed before, we obtained a set of matched impedances. Figure 5 shows the return loss, and Fig. 6

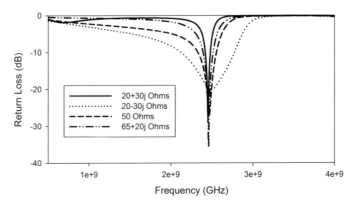

Fig. 5 Return Loss (dB) for different impedances at 2.45 GHz

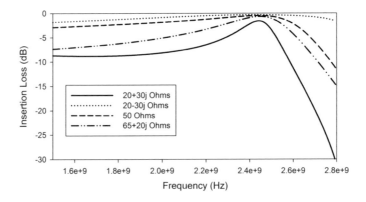

Fig. 6 Insertion Loss (dB) for different impedances at 2.45 GHz

Fig. 7 Smith chart coverage

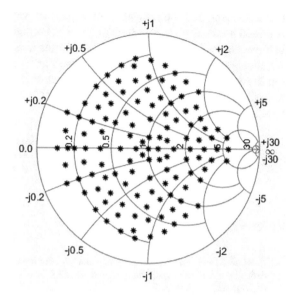

shows the insertion loss characteristics of the proposed impedance tuner for some of the matched load impedances such as 20+j30 ohms, 20-j30 ohms, 50 ohms, and 65+j20 ohms. It is seen from the graphs that for all the listed load impedances, the required design specifications are satisfied. Further, all the load impedances that could be possibly matched are plotted on the Smith chart as shown in Fig. 7. This gives an easy visualization of the wide tuning range of the proposed design.

A similar L-section varactor-based impedance tuner network using a linearized topology of anti-series varactor pair with anti-parallel varactor pair connection at its center tap is reported by Heebl et al. [20]. In comparison, the proposed design has lesser anti-series varactor pairs making it less bulkier and has a wider tuning range. Chen et al. [3] analyzed another adaptive impedance tuner for a broadband frequency range of 0.69–2.7 GHz. This circuit has a higher tuning range compared to that of the proposed design, however, it takes neither the non-linearities nor the distortions into account. Sánchez-Pérez et al. [21] has proposed another varactor-based tuner operating at 2.45 GHz with a double stub topology and using an anti-series varactor pair configuration to improve the power handling capability. But in comparison, this circuit has a lower tuner range when compared to that of the proposed design.

3   Conclusion and Future Work

In this paper, we have proposed a varactor-based impedance tuner network operating at an ISM frequency band centered at 2.45 GHz. With fewer tunable elements and control voltages, the proposed tuner circuit has a better tuning factor or in other words a better coverage of the possible load impedance values on the Smith chart. It has also used a linearized topology with better power handling capability. The idea of an impedance inverter helps to include a variable inductance functionality. In future, to validate the simulation results, we intend to realize the filter prototype. Also, non-linear and inter-modulation distortion analysis is required to have a quantitative analysis on how the adopted linearized topology has improved the linearity and power handling capability.

Acknowledgements The authors would like to acknowledge the University for granting the Senior Research Fellowship (SRF) and Kerala Science Council for Science, Technology, and Environment (KSCSTE) for supporting the work financially under the Engineering Technology Program (ETP).

References

1. Y. Sun, J. Moritz, X. Zhu, Adaptive impedance matching and antenna tuning for green software defined and cognitive radio, in *Proceedings IEEE International Midwest Symposium Circuits and Systems* (2011), pp. 1–4
2. P. Sjoblom, H. Sjoland, An adaptive impedance tuning CMOS circuit for ISM 2.4-GHz band. IEEE Trans. Circuits Syst. I **52**(6), 1115–1124 (2005)
3. N.J. Smith, C. Chen, J.L. Volakis, An improved topology for adaptive agile impedance tuners, in *IEEE Antennas and Wireless Propagation Letters*, vol. 12, pp. 92–95 (2013)
4. C.L. Palson, D. Das Krishna, J. Mathew, B.R. Jose, M. Ottavi, V. Gupta, Memristor based adaptive impedance and frequency tuning network, in *2018 13th International Conference on Design and Technology of Integrated Systems In Nanoscale Era (DTIS), Taormina* (2018), pp. 1–2
5. P. Ramachandran, Z.D. Milosavljevic, C. Beckman, Adaptive matching circuitry for compensation of finger effect on handset antennas, in *Proceedings 3rd European Conference on Antennas Propag. (EuCAP'2009), Mar. 23–27* (2009), pp. 801–804
6. W.C.E. Neo et al., Adaptive multi-band multi-mode power amplifier using integrated varactor-based tunable matching networks. IEEE J. Solid-State Circuits **41**(9), 2166–2176 (2006)
7. M.T. ElKhorassani, A. Palomares-Caballero, A. Alex-Amor, C. Segura-Gómez, P. Escobedo, J.F. Valenzuela-Valdés, P. Padilla, Electronically controllable phase shifter with progressive impedance transformation at K band. Appl. Sci. **9**(23), 5229 (2019)
8. H.M. Nemati, C. Fager, U. Gustavsson, R. Jos, H. Zirath, Design of varactor-based tunable matching networks for dynamic load modulation of high power amplifiers. IEEE Trans. Microw. Theory Tech. **57**(5), 1110–1118 (2009)
9. A. van Bezooijen, et al., A GSM/EDGE/WCDMA adaptive series-LC matching network using RF-MEMS switches. IEEE J. Solid-State Circuits **43**(10), 2259–2268 (2008). A. van Bezooijen, M.A. de Jongh, C. Chanlo, L.C.H. Ruijs, F. van Straten, R. Mahmoudi, A.H.M. van Roermund, A GSM/EDGE/WCDMA adaptive series-LC matchnig network using RFMEMS switches. IEEE J. Solid-State Circuits **43**(10), 2259–2268 (2008)
10. S. Fouladi, F. Domingue, N. Zahirovic, R.R. Mansour, Distributed MEMS tunable impedance-matching network based on suspended slow-wave structure fabricated in a standard CMOS technology. IEEE Trans. Microw. Theory Tech. **58**(4), 1056–1064 (2010)
11. C. Hoarau, N. Corrao, J. Arnould, P. Ferrari, P. Xavier, Complete design and measurement methodology for a tunable RF impedance-matching network. IEEE Trans. Microw. Theory Tech. **56**(11), 2620–2627 (2008)
12. C.E. McIntosh, R.D. Pollard, R.E. Miles, Novel MMIC source impedance tuners for on-wafer microwave noise-parameter measurements. IEEE Trans. Microw. Theory Tech. **47**(2), 125–131 (1999)

13. C.L. Palson, D.D. Krishna, B.R. Jose, J. Mathew, M. Ottavi, Memristor based planar tunable RF circuits. J. Circuit Syst. Comp. **28**, 1950225 (2019)
14. A. Fischer, F. Forsberg, M. Lapisa et al., Integrating MEMS and ICs. Microsyst. Nanoeng. **1**, 15005 (2015)
15. S. Hilliker, Avoiding pitfalls in varactor circuit designs. IEEE Trans. Consum. Electron. **CE-22**(3), 195–202 (1976). https://doi.org/10.1109/TCE.1976.266803
16. Skyworks Solutions, SMV123x series: hyperabrupt junction tuning varactors, datasheet (2012)
17. A. Othman, R. Barrak, M. Mabrouk, Varactors nonlinear effects in tunable RF filters, in *2016 5th International Conference on Multimedia Computing and Systems (ICMCS), Marrakech* (2016), pp. 427–432
18. K. Buisman et al., Varactor topologies for RF adaptivity with improved power handling and linearity. IEEE/MTT-S Int. Microw. Symp. **2007**, 319–322 (2007)
19. C. Huang et al., Ultra linear low-loss varactor diode configurations for adaptive RF systems. IEEE Trans. Microw. Theory Tech. **57**(1), 205–215 (2009)
20. J.D. Heebl, E.M. Thomas, R.P. Penno, A. Grbic, Comprehensive analysis and measurement of frequency-tuned and impedance-tuned wireless non-radiative power-transfer systems. IEEE Antennas Propag. Mag. **56**(5), 131–148 (2014)
21. C. Sánchez-Pérez, C.M. Andersson, K. Buisman, D. Kuylenstierna, N. Rorsman, C. Fager, Design and large-signal characterization of high-power varactor-based impedance tuners. IEEE Trans. Microw. Theory Tech. **66**(4), 1744–1753 (2018)
22. Y. Chen, R. Martens, R. Valkonen, D. Manteuffel, Evaluation of adaptive impedance tuning for reducing the form factor of handset antennas. IEEE Trans. Antennas Propag. **63**(2), 703–710 (2015)

Deadlock Avoidance in Torus NoC Applying Controlled Move via Wraparound Channels

Surajit Das and Chandan Karfa

Abstract Wraparound channels in Torus Network-on-Chip (NoC) help in reducing overall hop counts of a given traffic. The cyclic paths formed by wraparound channels make Torus NoC vulnerable to deadlock. Virtual channels (VC) or dedicated buffers are used in existing routing algorithms for Torus NoC to make them deadlock free. In this work, we present a deadlock avoidance approach in Torus without using VC or dedicated buffer. An *Arc Model* is proposed by restricting some movements via wraparound channels. Using different combinations from *Arc Model*, variants of algorithms are possible. As an application of *Arc Model*, a deadlock-free routing algorithm for Torus NoC is presented in this work without using any dedicated buffer or VC. Experimental results of that algorithm support its deadlock freedom and its effectiveness in saving hop counts.

Keywords Network-on-chip · Torus NoC · Wraparound channel · Deadlock avoidance

1 Introduction

Torus or k-ary n-cube NoC is widely used due to its high path diversity and topological symmetry with uniform node degree. This topology is largely used over the past 30 years in commercial supercomputer and data centre [1]. A 5×5 Torus NoC with router numbers and co-ordinates are shown in Fig. 1a. It is convenient to visualise Torus with help of Mesh NoC by connecting all boundary routers of a Mesh NoC to the corresponding opposite boundary routers. These additional connections help in

https://www.iitg.ac.in/ckarfa.

S. Das (✉) · C. Karfa
Indian Institute of Technology Guwahati, Guwahati 781039, Assam, India
e-mail: d.surajit@iitg.ac.in

C. Karfa
e-mail: ckarfa@iitg.ac.in

© The Author(s), under exclusive license to Springer Nature Singapore Pte Ltd. 2022
B. Mishra et al. (eds.), *Artificial Intelligence Driven Circuits and Systems*,
Lecture Notes in Electrical Engineering 811,
https://doi.org/10.1007/978-981-16-6940-8_8

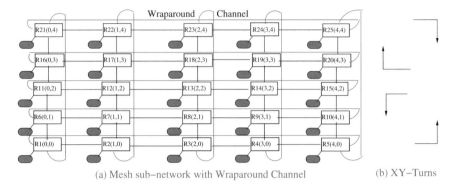

(a) Mesh sub–network with Wraparound Channel   (b) XY–Turns

Fig. 1 **a** 5 × 5 Torus NoC, **b** Permitted turns for XY-routing

reducing hop counts and are termed as wraparound channels. Nevertheless, they also create circular paths which make Torus prone to deadlock. While in some kinds of traffic a Torus may have better performance, simultaneously it is harder to achieve deadlock freedom and verify it in hardware. Therefore, research on deadlock detection and avoidance in Torus has importance. Deadlock arises due to circular dependency for resources that ultimately halts the system. For avoiding deadlock in Torus, different techniques like bubble flow control [2, 3] or virtual channels (VC) with datelines [4] are used. All these methods have overhead in terms of extra buffer or VCs. This work is aimed at avoiding deadlock in Torus without using extra buffer or extra channels inspired from two classical approaches: Turn Model [5] and Channel Dependency Graph (CDG) [6].

In this work, we visualise Torus NoC by logically subdividing it into Mesh sub-network and wraparound channels. Though wraparound channels are similar to other normal connections, we put a restricted move on wraparound channels. Marking some connections as wraparound channels are like dateline approach that is used for switching between VCs. In the dateline approach of deadlock avoidance, two VC classes VC0 and VC1 are used [4]. Certain co-ordinates are fixed for each X and Y-dimension for considering them as the dateline. If a packet is inserted into the network, it uses VC0. If a packet crosses the predefined dateline, for breaking the cyclic dependency, the packet is put from VC0 to another class VC1. Similarly, after crossing datelines (boundary of a Mesh NoC in our work), we put certain restrictions while forwarding a packet.

One approach of avoiding deadlock in Torus NoC or in ring network is to use extra buffers for implementing bubble flow control [2, 3]. Another approach is to use VCs [6–8] to overcome cyclic dependencies. However, these methods have overheads in terms of buffer usages, managing VCs and power consumption. Turn model (Fig. 1b) is an approach that does not use VCs or extra buffer for avoiding deadlock in Mesh NoC [5]. This motivates us to look into Turn model approach for deadlock avoidance in Torus. Deadlock verification works in [9] demonstrates that deadlock occurs by XY-routing in Torus NoC. Abstractly, the deadlock in Torus occurs because of the backward movement of a packet after taking wraparound channels in Torus. The question that motivates us: *Can we avoid deadlock in Torus without using VC and extra buffer and improve the utilisation of wraparound channels with certain restrictions in routing?* In this work, we, therefore, logically split a wraparound channel into *two arcs* to avoid a backward turn. We apply Turn model and take help of communicating finite state machine (CFSM)-based verification framework [9] for detecting deadlock while using those logically split wraparound channels. Since, the purpose of this work is to present a new deadlock avoidance approach only, rigorous performance evaluation and comparison of algorithms yielded from this approach is not presented in this work.

However, the *Turn model* is not applicable for deadlock detection for routing algorithms in Torus NoC. As per Turn Model, there is an impression that XY-routing is deadlock free in Torus NoC as well. On the other hand, XY-routing is deadlock prone in Torus NoC as shown in [9]. Therefore, it is desirable to have an accurate model to detect deadlock in Torus NoC. To the best of our knowledge, there is no such specific model available for accurate detection of deadlock in Torus NoC. Prime contributions of this work are

– An *Arc Model* is proposed for avoiding deadlock in Torus NoC without using extra buffer and VCs.
– As an utility of *Arc Model*, a deadlock-free deterministic routing algorithm for Torus NoC is demonstrated. More such algorithms are possible using *Arc Model*.
– The algorithm is implemented in CFSM based deadlock detection framework and compared with First Hop algorithm [5]. Experimental results show its effectiveness in saving hop counts.

The rest of the paper is organised as follows. Related works are presented in Sect. 2. Basic terminologies and analysis of deadlock for XY-routing in Torus NoC are presented in Sect. 3. The proposed *Arc Model* for deadlock avoidance in Torus NoC and its application is presented in Sect. 4. Experimental results are presented in Sect. 5. Finally, we conclude with the future direction of work in Sect. 6.

2 Related Work

Research on deadlock detection and avoidance for interconnection network and NoC is a more than three decades old problem with further scopes for improvements.

In bubble flow control, it uses dedicated buffer for avoiding deadlock in Torus NoC [2, 3]. It ensures at least one empty buffer slot in the ring that prevents deadlock. This approach has overhead in terms of maintaining an empty buffer. Dally et al. [6] introduce CDG for detecting the deadlock cycle. A necessary and sufficient condition for deadlock-free routing is the absence of a cycle in a channel dependency graph. The cycles in the channel dependency graph are removed by splitting physical channels into a group of VCs [6]. For eliminating the cycles, VCs are ordered and the routing algorithm is restricted to route packets in decreasing order. In dateline approach [4], two classes of VCs are used. A packet is forced to use another VC class after crossing the dateline to prevent cyclic dependency. Theorem on necessary and sufficient conditions for deadlock freedom of adaptive routing algorithms with VCs are presented in [8]. It provides an alternate escape path for messages that are involved in cyclic dependency.

Glass et al. [5] propose an alternate approach of designing deadlock-free routing algorithms called Turn model. Instead of adding extra buffer or VCs, the model is based on analysing the changes of direction by packets and the resultant cycles formed by them. All possible deadlock scenarios in Mesh NoC due to forbidden turns are well presented in these works [5]. For Torus NoC, use of wraparound channels are permitted only at the first hop (First Hop Algorithm) [5]. Otherwise, they result in deadlock. The restriction for using wraparound channel only at its first hop is eliminated in our work. For enhancing performance while using turn restrictions, research on effective turn distributions is presented in [10, 11]. Another deadlock avoidance approach partitions channels into disjoint sets without containing any cyclic dependency [12]. In that approach, VC is required for avoiding deadlock via wraparound channel. Since avoiding deadlock is expensive, in recent deadlock recovery research, different approaches are proposed for relieving from deadlock once it occurs [13–15].

To summarise, bubble flow control methods have overhead in terms of extra buffer [2, 3]. Similarly, VC methods have overhead for maintaining VCs and separate buffer corresponding to each VC [6, 8]. Turn model approach has no such overhead but wraparound channels are applicable only for limited packets that are injected only from boundary routers. This work intends for avoiding deadlock in Torus NoC with better utilisation of wraparound channel without using additional VC or buffer.

3   Deadlock in the Torus NoC and Dependency Graph

3.1   Deadlock in Torus NoC

Let us consider five packets p1(1, 3), p2(2, 4), p3(3, 5), p4(4, 1) and p5(5, 2) in a 5×5 Torus NoC in Fig. 2a. For each packet, the source and destination router numbers are put in the bracket. The current location of each packet is shown in Fig. 2a. For example, the packet p1(1, 3) has started from the router R1 is moving towards destination router R3 and is currently in the West port buffer of router R2. For the packet p4(4, 1), the shortest path $R4 \rightarrow R5 \rightarrow R1$ with the wraparound channel is going to be followed. In a similar way, the path for p2(2, 4) is $R2 \rightarrow R3 \rightarrow R4$, for p3(3, 5) is $R3 \rightarrow R4 \rightarrow R5$, and for p5(5, 2) is $R5 \rightarrow R1 \rightarrow R2$. Let us assume, each router has a buffer with a capacity of one packet, and packets p1, p2, p3, p4, and p5 are transmitted at the same time from their source routers. They have reached the input buffer of their next router, R2, R3, R4, R5, R1, respectively, as shown in

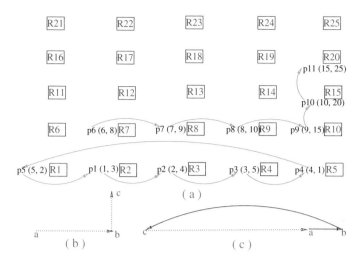

Fig. 2 **a** Buffer Dependency. **b** No Deadlock. **c** Deadlock

Fig. 2a. Therefore, the West input port Buffer of each router in Fig. 2a are full at the same moment. Packet p1(1, 3) is occupying the West buffer of router R2, waiting for the release of West port buffer of R3. Packet p2(2, 4) is occupying the same R3 buffer and is requesting the R4 buffer. Packet p3(3, 5) is occupying the R4 buffer and is requesting an R5 buffer. Packet p4(4, 1) is occupying the R5 buffer and is requesting an R1 buffer. Lastly, packet p5(5, 2) is occupying the R1 buffer and requesting an R2 buffer, currently occupied by packet p1. Thus, all five packets are waiting for each other for the release of buffer in a cyclic manner. This is an example of deadlock in Torus NoC. Though we have considered buffer capacity for only one packet in this example, deadlock is possible even if buffer capacity is for more than one packet but not unlimited.

3.2 Dependency Graph for Deadlock Representation

It is possible to spread buffer dependency in a direction between consecutive routers that have limited buffer, provided that each router generates packets destined in a specific direction, and destination nodes are at least two routers away from the source nodes. The exact path in which packets are waiting for the buffer is shown with a green arrow in Fig. 2a. *Buffer dependency graph represents only the direction in which dependency spreads without giving the router-wise detail path. It is like a CDG [6], as each vertex in the graph represents a buffer and its connecting input channel. One minor difference with CDG is, for representation convenience, all intermediate vertices are not put in the graph unless there is a change in direction.* The corresponding dependency graph for packet p6, p7, p8, p9, p10, is shown in Fig. 2b.

The corresponding dependency graph for packets p1, p2, p3, p4 and p5 is shown in Fig. 2c. Here, *dotted line* with arrow indicates the spreading of dependency without using a wraparound channel. *Solid* and *curved line* with arrow indicates the spreading of dependency due to the use of wraparound channels. Since the dependency graph forms a cycle in Fig. 2c, corresponding packets p1, p2, p3, p4 and p5 result in a deadlock. There is no deadlock for packet p6, p7, p8, p9, p10, as there is no cycle in Fig. 2b.

4 Deadlock Avoidance in Torus NoC

In a Torus NoC, direction wise there are four types of wraparound channels, namely, EW (from East to West boundary), WE (West to East), NS (North to South) and SN (South to North). Dependency graphs for wraparound channels are shown in Fig. 3. For the NS wraparound channel, there could be a sequence of packets waiting for each other's buffer in the cyclic path shown in Fig. 3a. The same case is applicable for the other three wraparound channels (Fig. 3b–d). Therefore, these wraparound channels are the primary cause of deadlock in Torus NoC.

As per the dependency graphs in Fig. 3, *backward movement immediately after taking any wraparound channel must be avoided to prevent deadlock*. After taking an X-wraparound channel EW, immediate movement *bd* towards the East has the potential for creating a deadlock in Fig. 3c. Similarly, after taking a Y-wraparound channel NS/SN, immediate movement in Y-direction towards the North/South has the potential for creating a deadlock. Therefore, restrictions need to be imposed for breaking the cycle created by wraparound channels. To achieve this, we propose an *Arc Model* in this work. *Arc Model* is applicable only to topologies like Torus where inherent cyclic paths (many rings) are present. Therefore, *Arc Model* is not applicable for NoC topologies like Mesh or Butterfly. Since no VC or additional buffer is used in *Arc Model*, there is no energy and area overhead in this approach as compared to the VC [7, 8] and bubble flow control [2, 3].

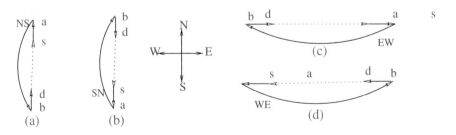

Fig. 3 Dependency for **a** NS, **b** SN, **c** EW and **d** WE wraparound channels

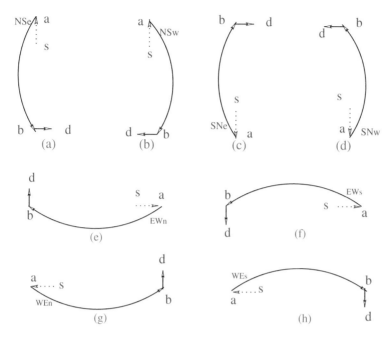

Fig. 4 Arc Model: **a** NSe, **b** NSw, **c** SNe, **d** SNw, **e** EWn, **f** EWs, **g** WEn, **h** WEs Arcs

4.1 Arc Model for Torus NoC

In our proposed Arc model, each wraparound channel is divided into two arcs by imposing restrictions on backward movement after the wraparound channel is taken. For example, the NS channel in Fig. 3a is divided into NSe (Fig. 4a) and NSw *arcs* (Fig. 4b). Essentially, after taking the NS channel, the packet is forced to move either in the East or West direction. Here, NSe /(NSw) *arc* is applicable if NS wraparound channel reduces hop counts and destination is in South-East/(South-West) direction. In a similar way, for breaking the cycle for all wraparound channels in Fig. 3, they are categorised into eight *arcs*. The eight possible *arcs* in the proposed *Arc Model* are shown in Fig. 4.

4.2 Application of Arc Model

For deadlock-free routing algorithm in Torus, instead of unrestricted wraparound channels, we use *arc(s)* from the *Arc Model* along with the XY-routing in the Mesh sub-network in such a way that they do not create cyclic dependency. Even though all the arcs are individually deadlock free, some of their combinations are deadlock prone. In this work, we have not systematically categorised all such arcs. More research with intensive verification is needed to categorise deadlock-free *arc* combinations, deadlock prone arc combinations and how many *arcs* can be used with a routing algorithm with deadlock freedom. In this work, we demonstrate one deadlock-free routing algorithm using two *arcs*. Variants of such algorithms using more *arcs* and more turns applying turn distribution [10, 11] are possible.

Algorithm 1: NE-SE Algorithm

Data: The source(S) and destination(D) co-ordinates of a packet in an NxN Torus NoC are S(x_s, y_s) and D(x_d, y_d), respectively. After each move, S is updated. Between a source and destination pair, the X-distance, $\Delta_x = |x_d - x_s|$ and the Y-distance, $\Delta_y = |y_d - y_s|$.

Result: The packet reaches destination when S == D.

1 **while** (($x_s \neq x_d$) $\vee$ ($y_s \neq y_d$)) **do**
2   **if** (($x_s < x_d$) $\wedge$ ($y_s > y_d$) $\wedge$ ($\Delta_y > N/2$)) **then**
3   NSe *arc* is applicable. Keeps on moving in the North direction. Once the North boundary is reached, move using the NS channel. Just after taking the NS wraparound channel, move one step East for the NSe *arc*. Follow XY-routing.
4   **else if** ($x_s < x_d$) $\wedge$ ($y_s < y_d$) $\wedge$ ($\Delta_y > N/2$) **then**
5   SNe *arc* is applicable. Keeps on moving in the South direction. Once the South boundary is reached, move using the SN channel. Just after taking the SN wraparound channel, move one step East for the SNe *arc*. Follow XY-routing.
6   **else**
7   Follow XY-routing.

The routing algorithm with (SNe + NSe) *arcs* with XY-routing for Torus NoC is presented as Algorithm 1. While comparing source and destination and calculating X-distance and Y-distance, we consider the co-ordinates in the same order as shown in Fig. 1. Since the packets that are destined towards North-East or South-East directions get advantage from using wraparound channels, we term this algorithm as *NE-SE algorithm*. Since all wraparound channels are not utilised, it is not a minimal routing algorithm for Torus. For the packet that uses *arc*, hop count is less than XY-routing and for the packet where *arc* is not applicable, hop count is equal to XY-routing. If an *arc* is applicable for a packet, first the corresponding wraparound channel is traversed. After that, XY-routing is used. The detailed steps are shown in Algorithm 1.

5 Experimental Results and Deadlock-Freedom Analysis

A communicating finite state machine (CFSM)-based simulation framework is used for all experiments in this work [9]. It detects confirmed deadlock with an exact deadlock scenario and reports overall hop counts saved. We consider three routing algorithms for experiments, (1) NSe + SNe + XY-turns (Algorithm 1), (2) EWs + WEn + XY-turns and (3) First Hop with XY-turns [5]. Algorithmic steps for the second algorithm are similar to Algorithm 1. Three types of traffic patterns are

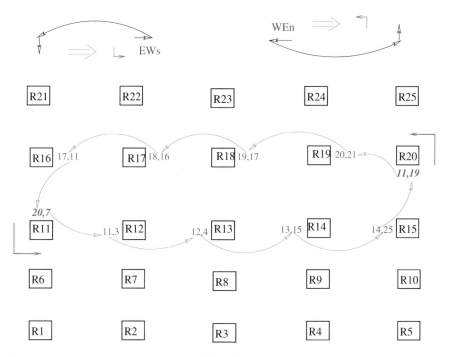

Fig. 5 Experimental Deadlock Snapshot: (EWs + WEn)

used in this work. Uniform traffic and Random permutation traffic are generated using Booksim simulator [16]. Another traffic pattern is generated using the random number function. In Experiment I, deadlock verification is performed. In Experiment II, saving of hop count is compared. All the experiments are performed in a single Intel Core i5 3.20 GHz, 8 GB RAM machine.

In Experiment I, deadlock is reported for the second algorithm (EWs + WEn + XY-turns) with all three types of traffic patterns. No deadlock is detected for the other two algorithms in any experiments performed with Torus NoC of gird size up to 12 × 12. An experimental deadlock snapshot for the second algorithm is depicted in Fig. 5. The CFSM based deadlock detector takes 2641 s to detect the deadlock in a 5 × 5 Torus NoC with random number traffic. An anti-clockwise cycle is formed using SE and NW turns added by (EWs + WEn) *arcs* along with XY-turns. Though SE and NW turns are YX-turns, they are unavoidable in boundary routers after using EWs or WEn *arcs*. The current position of a packet is represented using a number pair that indicates the source and destination of the packet. For example, a packet (17, 11) is currently arrived at East port of router R16 whose source and destination router is R17 and R11, respectively. The deadlock snapshot of the buffer dependency cycle is shown with a green arrow in Fig. 5. From Experiment I, it shows that some *arc* combinations are deadlock prone. A deadlock-free routing algorithms

Fig. 6 Deadlock Freedom:
(SNe + NSe + XY-Turns)

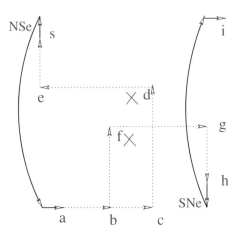

for Torus without using the dedicated buffer and VC are possible by careful use of *arc* combinations.

Deadlock-Freedom Analysis: CFSM-based framework detects confirmed deadlock. From a theoretical point of view, it does not guarantee the deadlock freedom if the deadlock is not found in a given traffic. Since no deadlock is detected in any experiment, it is likely that Algorithm 1 and First Hop Algorithm [5] are deadlock free. CFSM-based framework takes different times based on the size of the traffic pattern and NoC size to deliver all the packets at their destination router when deadlock is not detected. For the deadlock free (NSe + SNe + XY-Turns) it takes 4093 s to deliver a random traffic of 100000 packets in a 5 × 5 Torus NoC. An analysis for deadlock freedom is shown with help of Fig. 6. In Algorithm 1, SNe and NSe *arcs* do not create any extra turns besides XY-turns. Therefore, deadlock is not possible as per Turn model. For checking other possible cyclic dependencies involving *arcs*, a dependency graph is shown in Fig. 6. For an anticlockwise dependency cycle formation by NSe *arc*, a NW turn (not permitted) is needed at some vertex *d*. Similarly, for a clockwise dependency cycle formation by SNe *arc*, a SW turn is needed at some vertex, which is also not permitted. For spreading the dependency from NSe to SNe *arc*, a NE turn is needed at some vertex *f*, which is not permitted. Therefore, the dependency graph analysis shows that Algorithm 1 is deadlock free.

In Experiment II, Uniform (unf) and Random permutation (rpm) traffic with injection rate 0.05 and 0.08 are used for deadlock-free Algorithm 1 and First Hop Algorithm with XY-turns [5]. No deadlock is detected for any cases. For a packet with given source and destination, *Hop counts saved = (Manhattan distance between source and destination)* - (Actual distance traversed using an algorithm). Percentage of total hop counts saved in uniform and random permutation traffic are shown in Figs. 7 and 8, respectively. For First Hop Algorithm, hop count saving decreases with the increase of NoC size. Whereas, it does not very drastically for Algorithm 1. It shows better savings beyond 10 × 10 NoC (unf traffic) and 9 × 9 NoC (rpm traffic) in comparison to First Hop Algorithm.

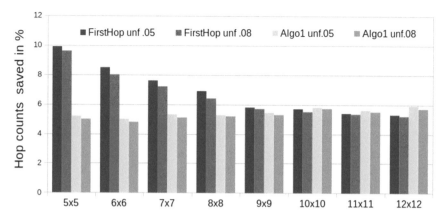

Fig. 7 First Hop Algorithm and Algorithm 1: Uniform Traffic

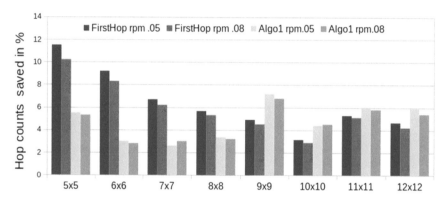

Fig. 8 First Hop Algorithm and Algorithm 1: Random Permutation Traffic

The Effect of the Ration r: The ratio $r = $ (*Number of routers present in boundaries/Total routers in a Torus NoC*), decreases as the NoC size grows. Therefore, the percentage of packets injected from boundary routers also decreases as the NoC size grows. In First Hop, wraparound channels are useful only for packets that are injected from boundary routers [5]. Therefore, the effectiveness of the First Hop Algorithm decreases as the NoC size grows. Whereas the r factor does not have adverse effects on *Arc model*-based algorithm. Algorithm 1 uses only two *Arcs*. Therefore, there is a possibility for more effective algorithms by applying more number of mutually deadlock-free *arcs* and turns for Torus NoC, irrespective of NoC size.

6 Conclusion and Future Work

Avoiding deadlock in Torus NoC without using dedicated buffer and VC is presented with *Arc Model* in this work. As an application, a routing algorithm is demonstrated. Deadlock freedom of that algorithm is shown using a buffer dependency graph. Experimental results detect no deadlock and show savings in hop counts. The algorithm is compared with a First Hop algorithm [5] that does not use any additional buffer or VC for deadlock avoidance as well. All eight *arcs* in *Arc Model* are individually deadlock free with XY-turns. Whereas, some *arcs* combinations are deadlock prone. Deadlock is detected for such an *arc* pair and an experimental deadlock snapshot is given. One future direction of our work is to systematically categorise all deadlock-prone *arcs* and deadlock-free *arcs* combinations with respect to given sets of turns besides XY-turns. It would help to develop enhanced algorithms using a different combination of *arcs* and turns. For better traffic distribution and better performance, applying turn distribution concept [10, 11] with *Arc Model* seems promising.

References

1. Z. Yu, D. Xiang, X. Wang, Vcbr: Virtual channel balanced routing in torus networks, in *2013 IEEE HPCC* (2013), pp. 1359–1365
2. V. Puente, C. Izu, R. Beivide, J.A. Gregorio, F. Vallejo, J.M. Prellezo, The adaptive bubble router. JPDC **61**(9), 111 (2001)
3. V. Puente, R. Beivide, J.A. Gregorio, J.M. Prellezo, J. Duato, C. Izu, Adaptive bubble router: a design to improve performance in torus networks, in *ICPP* (1999), pp. 58–67
4. W.J. Dally, B. Towles, *Principles and Practices of Interconnection Networks*, 1st edn. (Morgan Kaufmann, 2003)
5. C.J. Glass, L.M. Ni, The turn model for adaptive routing, in *[JACM*, vol. 41 (1994), pp. 874–902
6. Dally, Seitz, Deadlock-free message routing in multiprocessor interconnection networks. IEEE TC **C–36**(5), 547–553 (1987)
7. J. Duato, A new theory of deadlock-free adaptive multicast routing in wormhole networks, in *IPDPS* (1993), pp. 64–71
8. J. Duato, A necessary and sufficient condition for deadlock-free adaptive routing in wormhole networks. *IEEE TPDS* **6**(10) (1995)
9. S. Das, C. Karfa, S. Biswas, Formal modeling of network-on-chip using cfsm and its application in detecting deadlock. IEEE TVLSI **28**(4), 1016–1029 (2020)
10. Ge-Ming. Chiu, The odd-even turn model for adaptive routing. IEEE Trans. Parallel Distrib. Syst. **11**(7), 729–738 (2000)
11. B. Fu, Y. Han, J. Ma, H. Li, X. Li, An abacus turn model for time/space-efficient reconfigurable routing, in *ISCA* (2011), pp. 259–270
12. M. Ebrahimi, M. Daneshtalab, Ebda: a new theory on design and verification of deadlock-free interconnection networks, in *2017 ACM/IEEE 44th ISCA* (2017), pp. 703–715
13. M. Parasar, H. Farrokhbakht, N. Enright Jerger, P.V. Gratz, T. Krishna, J. San Miguel, Drain: deadlock removal for arbitrary irregular networks, in *HPCA* (2020), pp. 447–460
14. A. Ramrakhyani, P.V. Gratz, T. Krishna, Synchronized progress in interconnection networks (spin): a new theory for deadlock freedom, in *ISCA* (2018), pp. 699–711

15. M. Parasar, A. Sinha, T. Krishna, Brownian bubble router: enabling deadlock freedom via guaranteed forward progress, in *NOCS* (2018), pp. 1–8
16. N. Jiang, D.U. Becker, G. Michelogiannakis, J. Balfour, B. Towles, D.E. Shaw, J. Kim, W.J. Dally, A detailed and flexible cycle-accurate network-on-chip simulator, in *2013 IEEE ISPASS* (2013), pp. 86–96

Formal Modeling and Verification of Starvation Freedom in NoCs

Surajit Das and Chandan Karfa

Abstract Formal modeling plays a key role in verification of crucial properties of a complex system like Network-on-Chip (NoC) before the actual hardware is manufactured. This work presents detailed modeling of NoC routers using Finite State Machines (FSMs). We model buffer for storing packets, model switch for giving routing direction to packets, and model arbiter for resolving the conflict in case of packets from multiple input ports compete for the same output port. Synchronization among all these units and between adjacent routers are maintained in the proposed FSM-based model. As an application of the model, verification of starvation freedom is demonstrated for Fixed-priority and Round-robin arbitration logic. Thread level parallelism is used in experiments to speed up the verification process. The proposed model can be used to verify other properties in NoCs as well.

Keywords NoC · Formal model · Verification · FSM · Starvation

1 Introduction

Network-on-Chip (NoC) is an interconnection network between the cores in a Tiled Chip Multiprocessor (TCMP) [1]. A 3×3 mesh NoC with routers and connected processors are shown in Fig. 1a. A block diagram of a five-port NoC router is shown in Fig. 1b. The five bidirectional ports are namely East (E), West (W), North (N), South (S), and Local (L) ports. The L port is for the connection with the local processor. In an NoC, starvation, deadlock, and livelock are three major challenges that degrade the performance or lead to a situation where some packets never reach their destination

https://www.iitg.ac.in/ckarfa.

S. Das (✉) · C. Karfa
Indian Institute of Technology Guwahati, Guwahati 781039, Assam, India
e-mail: d.surajit@iitg.ac.in

C. Karfa
e-mail: ckarfa@iitg.ac.in

© The Author(s), under exclusive license to Springer Nature Singapore Pte Ltd. 2022   101
B. Mishra et al. (eds.), *Artificial Intelligence Driven Circuits and Systems*,
Lecture Notes in Electrical Engineering 811,
https://doi.org/10.1007/978-981-16-6940-8_9

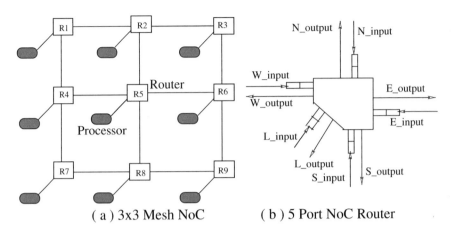

(a) 3x3 Mesh NoC   (b) 5 Port NoC Router

Fig. 1 **a** 3×3 Mesh NoC, **b** Input buffer NoC router with five ports

[1]. There is no inbuilt support for confirmed detection of these specifications using state-of-the-art NoC simulators like Booksim [2] or Gem5 [3]. Therefore, there is a need for formal modeling of NoC in detail level for the verification of such properties. NuSMV [4] is a verification tool that works using Finite State Machines (FSM)-based model. In this work, we present formal modeling of NoCs using FSMs for verification of properties with NuSMV. This work is not intended as a replacement for popular NoC simulators like Gem5 [3], Booksim [2] as they have many functionalities including performance analysis. Instead, in addition to such simulators, our proposed FSM-based model can be used for formal verification of various important properties of the NoCs in the pre-silicon stage.

Hermes NoC router architecture is followed in this modeling work [5, 6]. *Buffer*, *switch* and *arbiter* are three prime components for NoC functionality. *Buffers* are used as a temporary storage at an input port and the route computation of a packet is performed by *switch*. After route computation is done, the packet competes for the desired output port. In each output port, an *arbiter* is present that resolves the conflict by picking one packet from the competing packets. The selected packet is transmitted to the next router. Once a packet is moved to the next router, the buffer storage in the previous router needs to be cleared. Processing of the next packet starts only after transmitting the current packet. Therefore, we have to model the functionalities of buffer, switch and arbiter in each router using FSMs. In addition, modeling synchronization between buffer, switch, and arbiter and between two adjacent routers is important for smooth functioning and lossless packet transmission in NoC routers. In this work, the synchronization between the buffer, switch, and arbiter is maintained with two dedicated FSMs named *sync* and *return* in each port. The synchronization is maintained using the handshaking principle. High-level overview of packet flow between two routers and interaction between different units are shown in Fig. 2. The blue solid arrows show possible paths for the movement of packets. The dotted double-ended arrow indicates controlling of synchronization between components.

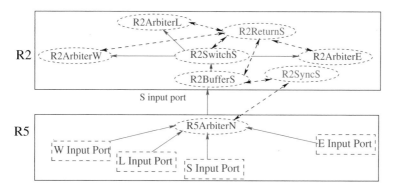

Fig. 2 Movement of packets (R5 → R2): synchronization overview

Table 1 Short form used for describing transitions in FSM diagrams

Functional unit	Short form	Functional unit	Short form
Arbiter	Ar	Buffer	Bu
Priority	Pr	Return	Re
Switch	Sw	Sync	Sy

While describing the FSM transitions, short forms (in Table 1) are used for each functional unit.

Starvation freedom ensures fairness in resource allocation. In NoC context, the output port and its connection to the next router acts as a resource of conflict. If multiple packets from different input ports try to access that resource continuously, there should be fairness such that competing packets from all input ports get a fair chance to be transmitted. Starvation freedom in NoC primarily depends upon underlying arbitration policies like Fixed-priority arbiter, Round-robin arbiter, Weighted Round-robin arbiter, latency based Slack-time arbiter, etc., controlling an output port of a router. As an application of our model, we target verification of starvation freedom considering Fixed-priority and Round-robin arbiters in this work. Specifically, the prime contribution of this work is summarized as follows:

– Formal modeling of router components like buffer, switch, and arbiter using FSM.
– Maintaining synchronization between them and designing Fixed-priority and Round-robin arbitration policies.
– As an application of the FSM-based model, verification of starvation freedom for Fixed-priority and Round-robin arbitration with NuSMV model checker.
– Verification time is significantly reduced by invoking parallel threads for individual routers.

Rest of the paper is organized as follows. Section 3 demonstrates the modeling of NoC using FSM. Representation of starvation specification using Linear Temporal

Logic (LTL) [7] for verification are described in Sect. 4. Experimental results and their analysis are put in Sect. 5. Related verification works on NoC are described in Sect. 2. Finally, we conclude in Sect. 6.

2 Related Works

There are a few works reported on formal modeling and verification of NoC targeting different properties. Due to its complexity, detailed modeling of NoCs is not performed in most of the works. In [8, 9], the communication infrastructure verification of NoC is targeted. A formal model for NoC is developed and implemented in the ACL2 theorem prover. The reachability of packets between routers is verified in these works. Detail modeling of individual NoC components via which packets pass through are not considered in these works [8–10]. Formal modeling of NoC using xMAS and progress between NoC components are shown in [11]. Verification of starvation freedom is not achieved in that work. Progress verification of a communication network is performed using xMAS-based model in [12]. A credit-based flow control is designed to control a number of packets that enter the system and to maintain synchronization. In our work, we maintain this flow control and synchronization using FSM-based model with help of a handshaking mechanism. A formal model of the Hermes NoC router architecture with its communication scheme is presented in [6]. Heterogeneous Protocol Automata (HPA) is proposed as a language to model Hermes NoC. The reachability of packets is verified in that work using Spin model checker [13]. We also consider Hermes NoC for modeling NoC in this work [5]. Hermes NoC router is a five bidirectional port router with a buffer at each input port. Verification of bidirectional NoC is performed in [14]. That work uses State Graph Manipulator (SGM) for model checking. Starvation freedom and mutual exclusion are verified in that work. For verifying starvation freedom, this work has not considered detailed resource allocation logic. In our work, we have modeled both Fixed-priority and Round-robin policy for starvation verification. In [15], process-algebra is used for formal verification of fault-tolerant NoC. Scalability is not achieved in that work as only 2×2 NoC model is considered for experiments. Scalability of NoC or lack of internal details of a router are the main issues in most of the NoC verification works. Detecting confirmed deadlock in NoC using a formal model is presented in [16]. Communicating finite state machine (CFSM) is used for modeling NoC in that work.

Verification of starvation freedom in NoC considering detail router design and detail resource allocation policy are not targeted in the existing works to the best of our knowledge. In this work, we target verification of starvation freedom considering Fixed-priority and Round-robin resource allocation policy using detail FSM-based model.

3   Formal Modeling of NoC

Modeling different functional units of NoC are performed using FSM in this work, as they are convenient to implement using NuSMV [4] model checker. Different FSM states corresponding to an NoC component represent its current state of operation. Logic for the next transition of an FSM depends upon the current states of its dependent functional units (FSMs). All transitions in FSMs are proceeded by handshaking with dependent FSMs. NuSMV internally composes all FSMs. Each functional unit is named with its associated router number and port name that are put as prefix and suffix, respectively. For example, the buffer present at the South (S) input port of router R2 is denoted by *R2BufferS*. Figure 1a is used to refer router numbers according to their positions in representing FSM models for different NoC functional units in this section.

3.1   Sync and Buffer

A *sync* FSM maintains synchronization between the *buffer* and the adjacent router from which the buffer accepts packet. We consider R2BufferS and R2SyncS for explaining the *sync* and *buffer* model. The FSM model for R2SyncS and all transitions are shown in Fig. 3a. Current state of $R2SyncS = 0(/1)$ means the R2BufferS contains zero(/one) packet. Current state of $R2SyncS = 01(/12)$ means the buffer needs to change its state to $R2BufferS = 1(/2)$. If $R2BufferS = 2$, it means the buffer is full. If $R2BufferS = 0(/1)$, it means the buffer has capacity to store two(/one) packets(/packet). The condition $(R2ReturnS = 0)$ indicates that S input port at R2 is ready to receive a packet. The condition $(R2ReturnS = 0)$ remains true until the packet is transmitted. The condition, $R5ArbiterN \neq Start$, indicates that $R5ArbiterN$, at N port of R5, has selected a packet for transmission into S

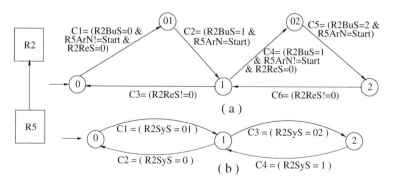

Fig. 3 a R2SyncS: Synchronization at the South port of router R2, **b** R2BufferS: Buffer at the South port of router R2

input port of R2. When the transmission is over, the arbiter returns to the initial state and the condition becomes $R5ArbiterN = Start$. The condition C1 in Fig. 3a indicates that R5ArbiterN is ready to transmit a packet to R2BufferS and the state of *R2SyncS* becomes *01*. The condition C2 indicates that R2BufferS receives the packet and R5ArbiterN returns to the initial state and the state of *R2SyncS* becomes *1*. R2BufferS uses the current states of R2SyncS FSM for its transitions.

The FSM for R2BufferS are shown in Fig. 3b. The condition C1 in Fig. 3b indicates that R2BufferS receives a packet and changes its state to *1*. Condition C2 indicates that the buffer becomes free. Satisfying condition C3 indicates that the buffer receives another packet.

3.2   Switch for Route Computation

A switch accepts packets from the buffer and diverts it to the desired output port for transmitting into the next router. The FSM representation of switch R2SwitchS is shown in Fig. 4. It remains in its initial state *Wt* (Wait) until no packet arrives in the buffer. When a packet arrives and the condition C1 is satisfied, the FSM changes its state to *C* (Compute). At the state *C*, routing direction is decided by a variable named *Route*. It takes any values randomly 0 or 1 or 2, indicating the direction toward L, E, and W port, respectively. Based on the routing direction, R2SwitchS reaches the appropriate state. Once the packet is transmitted, the respective condition C5(/C6/C7) becomes true and the FSM returns back to the initial state *Wt*. For developing a formal model- based simulation framework using the proposed NoC model, the routing direction of a packet would be decided by invoking a routing algorithm at state *C*.

Fig. 4 R2SwitchS: Switch at South port of router R2

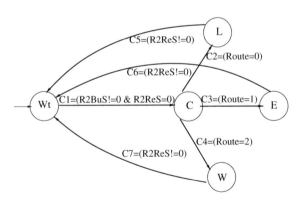

3.3   Returning to Initial State After a Transmission

After transmitting a packet, the corresponding *buffer*, *switch* and *arbiter* return to their initial states which are controlled by the *return* FSM. If the current state of the *return* FSM is other than 0, it indicates that the packet is transmitted and corresponding FSMs involved in that transmission must return to their initial states. After corresponding FSMs return to their initial state, the *return* FSM also changes to the initial state 0. An input port is ready for processing a new packet when the current state of the *return* FSM is 0.

The R2ReturnS at the S input port of router R2 is shown in Fig. 5. The initial state of R2ReturnS is 0. If R2BufferS contains one packet which is transmitted via E output port and condition C3 is satisfied, the state of the R2ReturnS changes to 1*E*. Here, condition C3 indicates that the S input port buffer of router R2 contains one packet, which is routed toward E output port by the switch and the E output port arbiter has selected the packet for transmission. The other condition C4 indicates that the S input port buffer of router R2 becomes empty after transmission of the packet, the S input port switch of router R2 returns to the initial state, and the E output port arbiter of router R2 returns to the initial state S (Start). Once corresponding FSMs related to this transmission return to their initial states, the condition C4 is satisfied and the R2ReturnS also returns to its initial state. Other transitions in Fig. 5 also progress in a similar way.

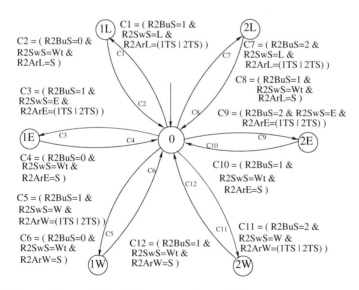

Fig. 5 R2ReturnS: Return FSM at the South port of router R2

3.4 Arbiter for Resolving Conflict at Output Port

Arbiter resolves the conflict between multiple packets from different input ports competing for the same output port. Some popular arbitration policies are Round-robin, Weighted Round-robin, First-come-first-serve, Fixed-priority, etc. In this work, we design an arbiter with Fixed-priority and Round-robin policy.

Fixed-Priority Arbiter In the Fixed-priority arbiter, priority is fixed. We choose the priority sequence of Local > East > West > North > South, i.e., packets from the L input port have the highest priority and the packet from the S input port has the lowest priority. The FSM representation of Fixed-priority arbiter (R2ArbiterS) present at the S port of a R2 is shown in Fig. 6. It transmits packet to the buffer R5BufferN in the adjacent router R5. Initially, the arbiter is in the initial state *Start*. If the buffer R5BufferN in the adjacent router R5 contains zero(/one) packet and the condition C1(/C7) is satisfied, a packet from L input port wins arbitration at R2ArbiterS. Here, the condition C1 indicates that the L input port of R2 is ready (R2ReL=0) for routing a packet which is destined toward the S output port (R2SwL=S) and the N input port buffer in the next router R5 is empty (R5BuN=0). Once C1(/C7) is satisfied, the arbiter changes its state to *1TL(/2TL)*. Once the transmission is over and the R5BufferN receives the packet and the condition C2(/C8) is satisfied. The condition C2 indicates that the L input port is under the process of synchronization (R2ReL!=0) after transmitting a packet and the packet is received at the N input port buffer of the next router R5 (R5BuN=1). When C2 is satisfied, the arbiter returns to its initial state *Start*. Other transitions in Fig. 6 also work in a similar way.

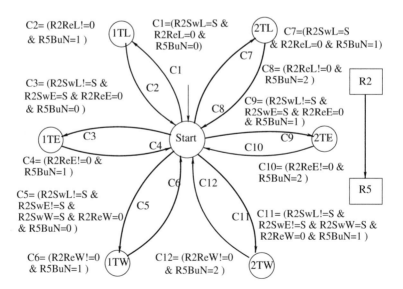

Fig. 6 R2ArbiterS: Fixed-priority arbiter at the South port of router R2

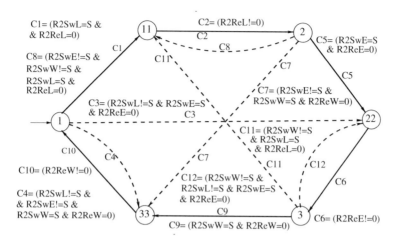

C1= (R2SwL=S &
& R2ReL=0)

C8= (R2SwE!=S &
R2SwW!=S &
R2SwL=S &
R2ReL=0)

C3= (R2SwL!=S & R2SwE=S
& R2ReE=0)   C3

C10= (R2ReW!=0)

C4= (R2SwL!=S &
& R2SwE!=S &
R2SwW=S & R2ReW=0)

C2= (R2ReL!=0)

C5= (R2SwE=S
& R2ReE=0)

C7= (R2SwE!=S &
R2SwW=S & R2ReW=0)

C11= (R2SwW!=S
& R2SwL=S
& R2ReL=0)

C12= (R2SwW!=S &
R2SwL!=S & R2SwE=S
& R2ReE=0)

C9= (R2SwW=S & R2ReW=0)

C6= (R2ReE!=0)

Fig. 7 R2PriorityS: Priority for Round-robin arbiter at the South port of router R2

Round-Robin Arbiter In Round-robin arbiter, the priority is not fixed but keeps on updating dynamically. After transmitting a packet from one port, the priority of that port is set to be lowest. It helps in facilitating a fair chance for the packets from other ports. For simplicity, two separate FSMs are used in Round-robin logic, one is Round-robin scheduler that sets priority dynamically and the other one is Round-robin arbiter that operates based on the current priority. Since priority is predefined in Fixed-priority arbiter, a single FSM is sufficient in case of a Fixed-priority arbiter. Figure 7 shows the Round-robin scheduler. Priority information from the Round-robin scheduler is used by the Round-robin arbiter. Initially, Round-robin scheduler is in the initial state *1*. The priorities 1, 2, 3 is corresponding to L, E, and W port, respectively. The priority 11(/22/33) indicates that a packet presents at the L (/E/W) port wins the arbitration. Transitions from Round-robin scheduler in Fig. 7 set priority to a packet from an input port in Round-robin fashion. If the current state is *1* and there is a packet from L input port and condition C1 is satisfied, the priority changes to *11* in Fig. 7. The condition C1 indicates that there is a packet from the L input port toward the S output port of router R2. Let there is a packet from E input port and there is no packet from L input port, and condition C3 is satisfied. The condition C3 indicates that there is no packet from the L input port toward the S output port, whereas a packet is destined toward the S output port from the E input port of router R2. If C3 is satisfied, the priority changes from state *1* to state *22*. Different priorities in Fig. 7 are set dynamically in a similar way. Round-robin arbiter operates using this priority information. The FSM states and transitions of Round-robin arbiter R2ArbiterS are shown in Fig. 8. The condition C1 in Fig. 8 indicates that the priority for the S output port arbiter at router R2 is set to 11 scheduling the L input port and the adjacent N input port buffer in router R5 is empty. The C7 represents the same condition as C1 except that the adjacent N input port buffer in R5 is storing one packet and it has the capacity for storing another packet. When the condition C1(/C7) is satisfied in

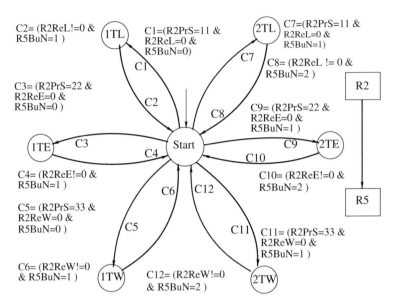

Fig. 8 R2ArbiterS: Round-robin arbiter at the South port of router R2

Fig. 8, the packet from L port gets a chance for transmission and the new state of the R2ArbiterS becomes *1TL(/2TL)*. Other transitions in the Round-robin arbiter are also carried out in a similar manner.

4 Starvation Specification Using Linear Temporal Logic

Starvation freedom ensures fairness in resource allocation. In NoC context, the output port and its corresponding buffer in the next router is a resource for conflict. If multiple packets from different input ports try to access that resource continuously, there should be fairness such that competing packets from all input ports get a fair chance to be transmitted. We encode our FSM based NoC models in NuSMV model checker [4] and give the starvation-freedom specification using linear temporal logic (LTL) [7] for verification. NuSMV internally composes all the FSMs and critically checks for all possible scenarios [7]. The model checker reports whether given LTL properties are satisfied or not.

LTL for Starvation Freedom When there is a packet competing for an output port, eventually that packet should be transmitted via that output port. Let there is a packet from W input port to S output port in router R2. That packet should get transmitted eventually via the S output port. This starvation-freedom property is expressed as, "When there is a packet from the W input port toward the S output port of router R2, the packet should get transmitted through the S output port eventually".

The LTL representation of that specification is, G ((R2ArbiterS=start)&
(R2SwitchW = S) ⟹ F ((R2ArbiterS = 1TW) ∨ (R2ArbiterS
= 2TW)). If this property is satisfied, it indicates that starvation freedom for W
input port at S output port of R2 is satisfied. But if this property fails, it indicates that
there are some possible scenarios where the packet from W input port to S output
port never gets transmitted. In router R1 of Fig. 1a, R1ArbiterE accepts packets from
L and S input ports. The starvation-freedom property for L input port at R1ArbiterE
is, "When there is a packet from the L input port toward the E output port of router
R1, the packet should get transmitted through the E output port eventually". Its LTL
representation is, G(((R1ArbiterE = Start) ∧ (R1SwitchL = E)) ⟹
F ((R1ArbiterE = 1TL) ∨ (R1ArbiterE = 2TL)). It means, globally when
the FSM R1ArbiterE is in state *Start* and FSM R1SwitchL is in state *E*, eventually
the state of R1ArbiterE will be *1TL* or *2TL*. The arbiter state *1TL* or *2TL* indicates
that the arbiter has selected packets from L input port. If this property is satisfied,
the starvation freedom for packets from L input port is satisfied at R1ArbiterE. In a
similar way, starvation freedom for other input ports are represented using LTL.

5 Experimental Results and Analysis

All experiments in this work are performed using an Intel Xenon(R) 2.10 GHz × 32
processors, 64 GB RAM machine. All routers in an NoC (from 2 × 2 to 8 × 8 mesh
NoC) are encoded with NuSMV as per presented FSM model. Since verification of
a big system like NoC is time-consuming process, they are executed using parallel
threads to reduce execution time. Since the starvation freedom is a locally dependent
property within a router, it is feasible to verify the starvation freedom for different
routers in parallel.

Verification of Starvation Freedom Verification results for starvation freedom at
arbiters of router R5 are shown in Table 2. For packets from L, E, W, N, and S input
ports, whether starvation-freedom specification is satisfied or not is denoted by T

Table 2 Starvation freedom for Fixed-Priority (FP) and Round-Robin (RR) arbiters

Input	Starvation freedom at arbiters									
	R5ArbiterL		R5ArbiterE		R5ArbiterW		R5ArbiterN		R5ArbiterS	
Port	FP A.	RR A.	FP A.	RR A.	FP A.	RR A.	FP A.	RR A.	FP A.	RR A.
L	NA	NA	T	T	T	T	T	T	T	T
E	T	T	NA	NA	T	T	T	T	T	T
W	T	T	T	T	NA	NA	F	T	F	T
N	F	T	F	T	F	T	NA	NA	F	T
S	F	T	F	T	F	T	F	T	NA	NA

Table 3 Comparing verification time for serial and parallel execution

Mesh	Serial Exe. (H:M:S)		Parallel Exe. (H:M:S)		Seep Up	
NoC	FP A.	RR A.	FP A.	RR A.	FP A.	RR A.
2 × 2	0.0.2	0.0.8	0.0.2	0.0.7	1x	1.1x
3 × 3	1.20.02	3.38.16	1.10.09	2.26.34	1.1x	1.3x
4 × 4	4.0.17	11.36.23	1.17.09	2.35.16	3.1x	4.5x
5 × 5	8.02.11	23.54.51	1.26.07	2.41.02	5.6x	8.9x
6 × 6	13.20.31	40.32.56	1.36.27	2.58.21	8.3x	13.7x
7 × 7	20.12.03	61.30.42	2.05.03	3.57.06	9.6x	15.6x
8 × 8	28.02.16	86.48.02	2.38.56	4.49.27	10.6x	18.0x

(true) or F (false). Starvation freedom is not applicable (NA) for a port if that port is inactive or there is no packet from that port.

For fixed priority (FP) arbiter, it seems obvious that starvation freedom is satisfied only for the highest priority port. On the other hand, experimental results in Table 2 shows that *starvation freedom is satisfied for both highest and second-highest priority ports*. The priority order is L > E > W > N > S. Packets from L input port have the highest and packets from S input port have the lowest priority. At the N output port of router R5 (R5ArbiterN), Table 2 shows that W and S input ports suffer from starvation, whereas L and E input ports are free from starvation for FP arbiter.

Debug trace of FP arbiter shows a scenario for continuous packets from L, E, W, S input ports competing for the N output port (R5ArbiterN). The packet from the highest priority port (L input port) is transferred to the next router. After transmitting a packet from an input port, there involves a delay in performing synchronization with the buffer (clearing buffer) and switch using the return FSM. Whereas, a packet from the next highest priority port (E input port) is ready and waiting for the same N output port. Therefore, the packet from E input port gets transmitted instead of transmitting another packet from the L input port. The synchronization latency is the reason for winning arbitration by a second-highest priority port as well in FP arbiter. By the time the E input port packet is transferred, the L input port is synchronized. Another packet from L input port wins arbitration for the next transmission. In this way, packets only from L and E input ports get transmitted alternatively via R5ArbiterN when there are continuous packets. Packets from W and S input ports suffer from starvation.

For Round-robin (RR) arbiter, experimental results in Table 2 show starvation freedom for all ports. In the RR arbiter, the priority is dynamically changing by the priority FSM shown in Fig. 7. Therefore, no input port suffers from starvation in the RR arbiter. Experimental results reinforce that our design and implementation are correct.

Run Time Improvement by Parallel Execution The execution time for starvation verification is shown in Table 3. We verify starvation freedom for each router in an NoC using parallel thread. Table 3 shows the execution time to verify starvation freedom for all routers in an NoC. State space increases due to the implementation of dynamic priority in Round-robin arbiter as compared to Fixed-priority arbiter. Therefore, verification time for NoC with a Round-robin arbiter increases significantly than that of NoC with a Fixed-priority arbiter. Table 3 shows that verification time increases with the increase of NoC sizes. In case of larger NoC, the number of routers increases. Therefore, verification time increases with the increase of NoC grid size. *In case of parallel execution, execution time is significantly saved for NoC with a higher grid size. Therefore, the parallelization of the verification process is very effective in the verification of starvation.* Table 3 shows that using parallel threads speed ups the verification process up to 18x times over it equivalent serial execution for bigger NoCs. The speed up is calculated as the ratio of serial execution time to the parallel execution time. Virtual Channels (VC) are not modeled in this work. Prime challenge with implementing VCs is the increased state space and verification time.

6 Conclusion and Future Work

This work presents formal modeling of NoC components using FSM in detail. Synchronization between different functional units are taken care for error-free functioning of the overall system. As an application of the presented model, verification of starvation freedom using Fixed-priority arbiter and Round-robin arbiter is demonstrated in this work. Since formal verification is a time-consuming procedure, thread-level parallelism is used in our experiments. Experiment results show significant improvement in reducing verification time over its equivalent serial execution. Other possible applications and future direction of work using the presented NoC model is verification of successful transfer of packets between routers and verification of progress in the whole NoC. For verification of properties like deadlock and livelock, all routers in the NoC system need to be considered at a time that leads to state-space explosion problem. Therefore, constructing a simulation framework with the proposed formal model for verification of globally dependent properties like deadlock and livelock is another future direction of work.

References

1. W.J. Dally, B. Towles, *Principles and Practices of Interconnection Networks*, 1st edn. (Morgan Kaufmann, 2003)
2. N. Jiang, D.U. Becker, G. Michelogiannakis, J. Balfour, B. Towles, D.E. Shaw, J. Kim, W.J. Dally, A detailed and flexible cycle-accurate network-on-chip simulator, in *2013 IEEE ISPASS* (2013), pp. 86–96
3. N. Binkert, B. Beckmann, G. Black, S.K. Reinhardt, A. Saidi, A. Basu, J. Hestness, D.R. Hower, T. Krishna, S. Sardashti, R. Sen, K. Sewell, M. Shoaib, N. Vaish, M.D. Hill, D.A. Wood, The gem5 simulator. SIGARCH Comput. Archit. News **39**(2), 1–7 (2011)

4. A. Cimatti, E.M. Clarke, F. Giunchiglia, M. Roveri, Nusmv: A new symbolic model verifier. CAV '99 (Springer, London, UK, UK, 1999), pp. 495–499
5. F. Moraes, N. Calazans, A. Mello, L. Möller, L. Ost, Hermes: an infrastructure for low area overhead packet-switching networks on chip. Integration **38**(1), 69–93 (2004)
6. V.A. Palaniveloo, A. Sowmya, Application of formal methods for system-level verification of network on chip, in *2011 IEEE ISVLSI* (2011), pp. 162–169
7. C. Baier, J.-P. Katoen, *Principles of Model Checking (Representation and Mind Series)* (The MIT Press, 2008)
8. D. Borrione, A. Helmy, L. Pierre, J. Schmaltz, A generic model for formally verifying noc communication architectures: a case study, in *NOCS'07* (2007), pp. 127–136
9. D. Borrione, A. Helmy, L. Pierre, J. Schmaltz, Executable formal specification and validation of noc communication infrastructures, in *Symposium on Integrated Circuits and System Design* (ACM, 2008), pp. 176–181
10. S. El-Ashry, M. Khamis, H. Ibrahim, A. Shalaby, M. Abdelsalam, M.W. El-Kharashi, On error injection for noc platforms: a uvm-based generic verification environment. IEEE Trans. Comput.-Aided Des. Integr. Circuits Syst. **39**(5), 1137–1150 (2020)
11. S. Das, C. Karfa, S. Biswas, xmas based accurate modeling and progress verification of nocs, in *VLSI Design and Test* (Springer Singapore, Singapore, 2017), pp. 792–804
12. S. Ray, R.K. Brayton, Scalable progress verification in credit-based flow-control systems, in *DATE* (2012), pp. 905–910
13. Gerard J. Holzmann, The model checker spin. IEEE Trans. Softw. Eng. **23**(5), 279–295 (1997)
14. Y. Chen, W. Su, P. Hsiung, Y. Lan, Y. Hu, S. Chen, Formal modeling and verification for network-on-chip, in *2010 ICGCS* (2010), pp. 299–304
15. Z. Zhang, *Verification Methodologies for Fault-Tolerant Network-On-Chip Systems*. Ph.D. Dissertation, Department of Electrical and Computer Engineering, The University of Utah, USA (2016)
16. S. Das, C. Karfa, S. Biswas, Formal modeling of network-on-chip using cfsm and its application in detecting deadlock. IEEE TVLSI **28**(4), 1016–1029 (2020)

Reduced Graphene Oxide Soil Moisture Sensor with Improved Stability and Testing on Vadose Zone Soils

Kamlesh S. Patle, Salman siddiqui, Hemen K. Kalita, and Vinay S. Palaparthy

Abstract The vadose zone is the unsaturated soil surface that ranges from the earth surface to the groundwater table. Availability of the soil moisture in the vadose zone is crucial for the proper crop development and will lead to precise irrigation. Soil moisture sensors are extensively reported to measure the water content in the vadose zone. This work focuses on demonstrating the soil moisture micro-sensor, where reduced graphene oxide (rGO) is used as the sensing film. rGO is synthesized using the annealing of the graphene oxide sheets at 200 °C for 5 min. As-prepared rGO is drop casted on the inter-digitated electrodes with feature size of about 1500 μm $\times$ 2400 μm. The fabricated micro-sensors response is studied on the loamy sand soil collected from the agriculture field. Lab measurements show the sensor resistance varies from around 101 to 3 MΩ, when gravimetric water content (GWC) varies from 2 to 12%. The sensors response time is about 60–75 s, and the sensor output is stable for 5 months.

Keywords reduced Graphene Oxide (rGO) · Soil moisture sensors · Precise irrigation and vadose zone soils

K. S. Patle (✉) · V. S. Palaparthy
Department of Information and Communication Technology, DAIICT, Gandhinagar, India
e-mail: kamlesh_patle@daiict.ac.in

V. S. Palaparthy
e-mail: vinay_shrinivas@daiict.ac.in

S. siddiqui
Department of Physics, Indian Institute of Technology Bombay, Mumbai, India

H. K. Kalita
Department of Physics, Gauhati University, Guwahati, India
e-mail: hemenkalita@gauhati.ac.in

© The Author(s), under exclusive license to Springer Nature Singapore Pte Ltd. 2022
B. Mishra et al. (eds.), *Artificial Intelligence Driven Circuits and Systems*,
Lecture Notes in Electrical Engineering 811,
https://doi.org/10.1007/978-981-16-6940-8_10

1 Introduction

For the vadose zone (unsaturated soil), soil water content monitoring at regular intervals is required to improve the proper yield of crops, where soil moisture sensors play a vital role [1, 2]. Soil moisture in the vadose zone (unsaturated zone) is a key factor for an precise irrigation. For precise irrigation, prior information about soil in the vadose zone would be beneficial, considering the water conservation. For irrigation management systems soil moisture sensors are widely reported. Soil moisture can be measured by using the standard lab method and in-situ methods. Lab technique comprises standard gravimetric method and calcium carbide test. The disadvantage of the lab technique is that it is time consuming process and also requires the soil sampling from the field [3]. In-situ method comprises frequency domain reflectometry (FDR), time domain reflectometry (TDR), heat-pulse method, tensiometric, resistive method, etc. However, TDR and FDR sensing methodologies are expensive and not affordable for the farmers in developing countries [4, 5]. On the other hand, heat-pulse method, resistive and tensiometric methods are affordable but need frequent soil calibration and stability of these sensors is a concern [6–8].

Micro-sensor platforms, has a dimension of about micrometer in scale, have emerged as important tools for the agriculture sensing applications considering the price and accuracy. The rapid advancement of the micro-fabrication technology and micro-sensors has opened avenues for the large-scale sensor deployments in the field at an affordable price. Researchers are widely focused on using the micro-sensors for the in-situ agriculture applications and explored the potential uses of the micro-sensors for sensing soil moisture [9], pH [10], pathogens [11], nutrients [12] in soils. In micro-sensor platform, sensing films plays a vital role to enhance the sensor transfer characteristics such as sensitivity, selectivity, response time, hysteresis, etc. Researchers have explored various sensing films for soil moisture sensing such as PEDOOT:PSS polymers, SU8-Cabron black nano-composites, water soluble polymers, graphene derivatives etc. [13–16]. Out of the reported sensing films for agriculture applications, graphene derivatives found to be potential materials considering the sensitivity and stability, which is important for in-situ sensors deployed in the field [15, 16].

In this work, we explored the new graphene derivative, viz., reduced graphene oxide (rGO) and its soil moisture sensing properties. Application of rGO is widely reported for the humidity sensing [17–19]. However, the potential use of the rGO and its soil moisture sensing properties needs to be explored. In this study, rGO is selected considering its low-cost synthesis process and excellent stability with respect to time. We synthesized the rGO using the annealing of the graphene oxide sheets at 200 °C for 5 min. Further, we studied the fabricated micro-sensor transfer characteristics on the loamy sand soil collected from the agriculture field. Further we have explained the plausible sensing mechanism of soil moisture interaction with rGO.

2 Material and Methods

A *Reduced graphene oxide synthesis*

First, graphene oxide (GO) is synthesized using the modified Hummer's process as reported in [16]. Further, a solution is made, which comprises graphite (1 gm), $NaNO_3$ and H_2SO_4 with 0.76 gm and 62 gm, respectively, and then as-prepared solution is stirred in an ice bath at about 500–600 rpm. Furthermore, the prepared solution is taken out from the ice bath and then 0.5 gm of $KMnO_4$ is added to this solution and then cooled for 2 h and stir at 300 rpm for 5 days. Then 100 ml of H_2SO_4 (5 wt%) is added to this aqueous solution and kept for about 1 h followed by stirring of about 2 h. Then, H_2O_2 (30 wt% aqueous solution) with 3 gm in weight is added to the solution and followed by stirring for 2 h. Further, the purification of the prepared solution is done 15 times keeping the centrifugation constant at 5000 rpm for 5 min and adding mixed aqueous solution of H_2SO_4 (3 wt%) and H_2O_2 (0.5 wt%). This step will help to remove the oxidant ions from the solution followed by cleaning of solution with de-ionized water. Then, a highly dispersed GO solution is achieved. Then, aqueous solution of graphene oxide is drop casted on the Si/SiO_2 substrate and placed under the ambiance condition for 24 h. Then, substrates are placed in the glass tray and inserted in the furnace. Then, GO is exposed at 200 °C for 5 min under argon gas.

B *Surface characterization*

Figure 1a shows the AFM images of the rGO on the mica substrate. First The rGO on mica sheet is kept for 4 h for air drying. From Fig. 1b, it can be seen that a monolayer rGO sheet has thickness of about ±2 nm, which is in agreement with [16]. Further, Figure 1c shows the Raman spectrum of the as-prepared rGO where the D band at 1346 cm^{-1} and G band at 1595 cm^{-1}, it is attributed to the E2g phonons and sp2 carbon atoms [15, 16]. The I_D/I_G ratio (1.25) is achieved in the rGO, which confirms the defects present on the rGO film sample. Furthermore, to analyze the functional groups at the rGO surface, we did the X-ray photoelectron spectroscopy (XPS). From the wide XPS survey it is evident that the C1s peak at 285 eV represent the main graphite and an O1s peak at 534 eV confirm the presence of functional groups comprising the oxygen [16].

C *Micro-sensor and Experimental setup*

In this work, we have fabricated the micro-sensors inter-digitated electrodes (IDEs) using micro-fabrication methodology, which his reported in our earlier work [15]. The dimensions of the fabricated IDEs are about 1500 μm × 2400 μm. Fabricated micro-sensor is mounted on the printed circuit board (PCB) and two leads are taken out for the electrical measurements. Figure 2b shows the packaging of the micro-sensors with three layers of nylon mesh placed orthogonally and has a size of about 500 μ. Sensor packaging is important and crucial for the lab and in-situ measurements considering the price and reusability. The nylon mesh mounted on the sensor avoids the direct interaction of the soil particles with the fabricated sensors. Figure 2b shows the developed micro-sensor

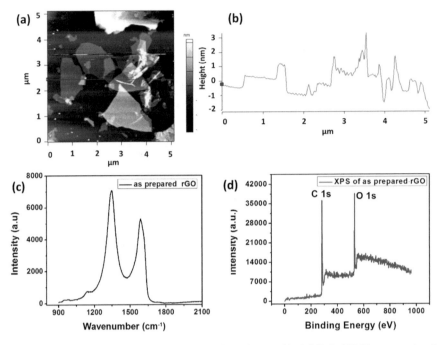

Fig. 1 **a** AFM image of the as-prepared reduced graphene oxide (rGO), **b** AFM image sectional analysis of rGO, **c** Raman spectrum of **d** Wide region XPS survey spectrum of as-prepared rGO sheets

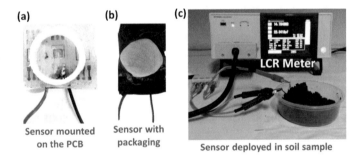

Fig. 2 **a** Fabricated micro-sensor placed on the PCB, **b** Micro-sensor packaging with protective three layers of nylon mesh, **c** developed sensor deployed in the loamy sand soil and response is recorded using the LCR Meter

inserted in the soil sample. Figure 2c shows the sensor is inserted in the soil sample and LCR meter records the sensor's output change in the resistance.

In this work, we used the vadose zone soil (unsaturated loamy sand soil), which has the saturation capacity of about 40%, field capacity is around 10% and permanent wilting point is 4%. Soil samples are prepared, and the developed sensor is inserted in the mold containing the soil sample with different soil water levels and corresponding changes in the sensor resistance are recorded. Details of the soil sample preparation are reported in our earlier work [20]. For the lab measurements the temperature and humidity are maintained constant at 25 °C and 55% RH, respectively. For the LCR meter, an excitation frequency and voltage are 1 V and 1 kHz, respectively.

3 Results and Discussion

Figure 3a shows the fabricated sensor response w.r.t to various gravimetric water content for loamy sand soil. From Fig. 3a it is apparent that when the gravimetric water content increases, the sensor resistance decreases. The resistance offered by the sensor is 101.1 MΩ, 72.3 MΩ, 13.94 MΩ, 7.64 MΩ, and 3.35 MΩ corresponding to 1, 3, 6, 9, and 12% gravimetric water content, respectively. This response observed in the sensors is due to the adsorption and interaction of water molecules present in the soil surface onto the reduced graphene oxide film. Soil moisture sensing mechanism

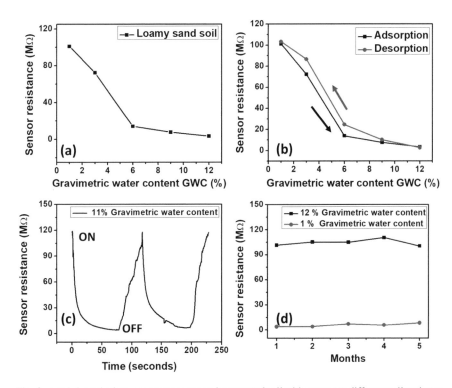

Fig. 3 **a** Fabricated micro-sensor response on loamy sand soil with respect to different soil moisture values, **b** hysteresis of the developed sensor on loamy sand soil, **c** Response time of the fabricated micro-sensor, **d** stability of the micro-sensor examined for 5 months

of rGO is explained further in this section. Figure 3b shows the hysteresis offered by the sensors. For hysteresis study, first the sensors are deployed in increasing order of gravimetric water content named as adsorption in Fig. 3b. Then, the developed sensor is inserted in the soil sample with a subsequent decrease in the soil moisture. From Fig. 3b it is evident that sensor exhibits a hysteresis of about ± 3% gravimetric water content. Hysteresis in the sensor response is recognized as the charge traps in the rGO film [15]. As shown in Fig. 3a, b, the flattening of the sensor response is attributed to bulk formation of hydrogen bonding in the second physisorbed layer.

In this work, we studied the response time of the developed micro-sensor as shown in Fig. 3c. For response time study, we maintained the gravimetric water content around 3% and then a fabricated sensor is inserted in the soil sample. From Fig. 3c is observed that sensor resistance becomes flat after 120 s (termed as on in Fig. 3c). Then sensors are removed from the soil and used air gun to remove the soil particles from the mesh. Thus, it is inferred that the sensor recovery time is about 30 s (termed as off in Fig. 3c). Further, we studied the stability of the sensors with respect to time, which one of the important factors for the soil moisture sensors deployed in the field. For stability study, we examined the sensor response for 5 months at two different water content as shown in Fig. 3d. We found the maximum standard deviation of about 3.7 MΩ and 1.5 MΩ for 12 and 1% gravimetric water content, respectively, in sensor resistance.

Figure 4 shows the schematic representation of the probable sensing mechanism of the soil moisture with rGO film. As shown in Fig. 4 at low soil water content

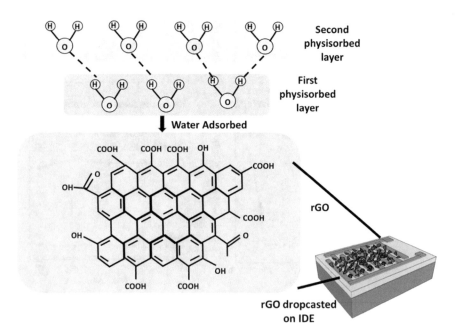

Fig. 4 Interaction of the soil moisture with the rGO film

Table 1 Comparison of the different sensing elements used for the soil water content measurements

Ref	Sensing material and platform	Response	Response time (S)	Stability (Months)	Hysteresis (% GWC)
15	Graphene quantum dots on IDEs	0.09×10^{-6} mho/1% for the white clay	120–180	4	–
16	Graphene oxide on IDEs	340% for 2.35% to 49% in the red soil	100–120	4	±2%
20	Molybdenum disulfide on IDEs	540% was obtained in black soil	65	–	±5%
13	Poly-aniline (PANI) nanofibers	275 μV/0.1% for, white clay	400–800	–	–
This work	Reduced graphene oxide	96% for loamy sand soil	60–75	5	±3%

the oxygen functional groups of the rGO surface through double hydrogen bonding physisorbed to the water molecules (named as first physisorbed layer in Fig. 4). Further, with increase in the soil water content the physical adsorption of water molecules becomes apparent and the water molecule available in second physisorbed layer binds to the first physisorbed layer water molecule via single hydrogen bonding [15, 16, 21].

Table 1 tabulates a novelty of the current work with various sensing material reported for the soil moisture measurements. From Table 1, it is seen that rGO offers the stability of around 5 months and hysteresis of about ±3 GWC. Response of the sensor is around 96% when the GWC varies from 1 to 12%. From Table 1, it can be concluded that rGO is one of the potential candidates for the in-situ field measurements.

4 Conclusions

For the vadose zone, monitoring the soil water content is mandated to enhance the yield of crops, where soil moisture sensors a considered to be important. Thus, in this work, we have fabricated the soil moisture micro-sensors, where rGO is used as the sensing film. Fabricated micro-sensors have sensitivity of about 96% for the loamy sand soil when soil water content varies from 1 to 12%. We observed that sensor response time is about 60–75 and has hysteresis around ±3% GWC. The sensor has a stability of around 5 months with respect to time. Thus, rGO finds to be a potential candidate for making the low-cost and stable soil moisture sensor for the in-situ measurements. Proposed rGO along with the IDEs can be further explored for different bio-sensing and explosive, environmental monitoring applications.

References

1. S.L. Su, D.N. Singh, M.S. Baghini, A critical review of soil moisture measurement. Measurement **54**, 92–105 (2014)
2. V.S. Palaparthy, M.S. Baghini, D.N. Singh, Review of polymer-based sensors for agriculture-related applications. Emerg. Mater. Res. **2**(4), 166–180 (2013)
3. F.S. Zazueta, J. Xin, "*Soil Moisture Sensors*", *Florida Cooperative Extension Service, Bulletin 292* (Institute of Food and Agricultural Sciences, University of Florida, Gainesville, 1994), pp. 1–11
4. G.C. Topp, J.L. Davis, A.P. Annan, Electromagnetic determination of soil water content: measurements in coaxial transmission lines. Water Resour. Res. **16**(3), 574–582 (1980)
5. B.H. Rao, D.N. Singh, Moisture content determination by TDR and capacitance techniques: a comparitive study. Int. J. Earth Sci. Eng. **4**(6), 132–137 (2011)
6. G. Pramod, M.C. Devendrachari, R. Thimmappa, B. Paswan, A.R. Kottaichamy, H.M.N. Kotresh, M.O. Thotiyl, Galvanic cell type sensor for soil moisture analysis. Anal. Chem. **87**(14), 7439–7445 (2015)
7. V.S. Palaparthy, S. Sarik, A. Mehta, K.K. Singh, M.S. Baghini, An automated, self sustained soil moisture measurement system using low power dual probe heat pulse (DPHP) Sensor, in *2015 IEEE Sensors* (IEEE, 2015), pp. 1–4
8. P. Aravind, M. Gurav, A. Mehta, R. Shelar, J. John, V.S. Palaparthy, K.K. Singh, S. Sarik, M.S. Baghini,. A wireless multi-sensor system for soil moisture measurement, in *2015 IEEE Sensors* (IEEE, 2015), pp. 1–4
9. J. Liu, M. Agarwal, K. Varahramyan, E.S. Berney IV, W.D. Hodo, Polymer-based microsensor for soil moisture measurement. Sens. Actuators B: Chem. **129**(2), 599–604 (2008)
10. S. Patil, H. Ghadi, N. Ramgir, A. Adhikari, V. Ramgopal Rao, Monitoring soil pH variation using Polyaniline/SU-8 composite film based conductometric microsensor. Sens. Actuators B: Chem. **286**, 583–590 (2019)
11. R.S. Patkar, M. Vinchurkar, M. Ashwin, A. Adami, F. Giacomozzi, L. Lorenzelli, M.S. Baghini, V. Ramgopal Rao, Microcantilever based dual mode biosensor for agricultural applications. IEEE Sens. J. **20**(13), 6826–6832 (2019)
12. R.S. Patkar, M.G. Seelan, V. Ramgopal Rao, A highly sensitive piezoresistive cantilever based sensor platform for detection of macronutrients in soil, in *2015 IEEE 15th International Conference on Nanotechnology (IEEE-NANO)* (IEEE, 2015), pp. 751–754
13. S.J. Patil, A. Adhikari, M.S. Baghini, V. Ramgopal Rao, An ultra-sensitive piezoresistive polymer nano-composite microcantilever platform for humidity and soil moisture detection. Sens. Actuators B: Chem. **203** (2014), 165–173
14. T. Jackson, K. Mansfield, M. Saafi, T. Colman, P. Romine, Measuring soil temperature and moisture using wireless MEMS sensors. Measurement **41**(4), 381–390 (2008)
15. H. Kalita, V.S. Palaparthy, M.S. Baghini, M. Aslam, Graphene quantum dot soil moisture sensor. Sens. Actuators B: Chem. **233**, 582–590 (2016)
16. V.S. Palaparthy, H. Kalita, S.G. Surya, M.S. Baghini, M. Aslam, Graphene oxide based soil moisture microsensor for in situ agriculture applications. Sens. Actuators B: Chem. **273**, 1660–1669 (2018)
17. S.Y. Park, Y.H. Kim, S.Y. Lee, W.S., J.E. Lee, Y.-S. Shim, K.C. Kwon et al., Highly selective and sensitive chemoresistive humidity sensors based on rGO/MoS 2 van der Waals composites. J. Mater. Chem. A **6**(12), 5016–5024 (2018)
18. H. Yang, Q. Ye, R. Zeng, J. Zhang, L. Yue, M. Xu, Z.-J. Qiu, D. Wu, Stable and fast-response capacitive humidity sensors based on a ZnO nanopowder/PVP-RGO multilayer. Sensors **17**(10), 2415 (2017)

19. S. Wang, G. Xie, Y. Su, L. Su, Q. Zhang, H. Du, H. Tai, Y. Jiang, Reduced graphene oxide-polyethylene oxide composite films for humidity sensing via quartz crystal microbalance. Sens. Actuators B: Chem. **255**, 2203–2210 (2018)
20. S.G. Surya, S. Yuvaraja, E. Varrla, M.S. Baghini, V.S. Palaparthy, K.N. Salama, An in-field integrated capacitive sensor for rapid detection and quantification of soil moisture. Sens. Actuators B: Chem. **321**, 128542 (2020)
21. H. Bi et al., Ultrahigh humidity sensitivity of graphene oxide. Sci. Rep. **3**(1), 1–7 (2013)

NeuralDoc-Automating Code Translation Using Machine Learning

Sai Sree Harsha, Aditya Chandrashekhar Sohoni, and K. Chandrasekaran

Abstract Source code documentation is the process of writing concise, natural language descriptions of how the source code behaves during run time. In this work, we propose a novel approach called NeuralDoc, for automating source code documentation using machine learning techniques. We model automatic code documentation as a language translation task, where the source code serves as the input sequence, which is translated by the machine learning model to natural language sentences depicting the functionality of the program. The machine learning model that we use is the Transformer, which leverages the self-attention and multi-headed attention features to effectively capture long-range dependencies and has been shown to perform well on a range of natural language processing tasks. We integrate the copy attention mechanism and incorporate the use of BERT, which is a pre-training technique into the basic Transformer architecture to create a novel approach for automating code documentation. We build an intuitive interface for users to interact with our models and deploy our system as a web application. We carry out experiments on two datasets consisting of Java and Python source programs and their documentation, to demonstrate the effectiveness of our proposed method.

Keywords Program comprehension · Automatic documentation · Neural machine translation · Transformer · BERT · Software engineering

1 Introduction

Programming is an integral part of software development. Software developers write programs in several iterations, and through active collaboration between teams, during which the program is frequently updated and optimised. Studies have shown

S. Sree Harsha (✉) · A. C. Sohoni · K. Chandrasekaran
Department of Computer Science and Engineering, National Institute of Technology, Karnataka, Surathkal, India

K. Chandrasekaran
e-mail: kchnitk@ieee.org

© The Author(s), under exclusive license to Springer Nature Singapore Pte Ltd. 2022
B. Mishra et al. (eds.), *Artificial Intelligence Driven Circuits and Systems*,
Lecture Notes in Electrical Engineering 811,
https://doi.org/10.1007/978-981-16-6940-8_11

that developers spend about 59% of their time in program comprehension activities [1], indicating that poorly documented source code could result in inefficiency and increased cost for the organisation. In the absence of documentation, programmers will be forced to painstakingly navigate through the source code to grasp its functionality. However, constructive and accurate documentation will assist programmers to successfully innovate over existing code resulting in better solutions. In most cases, maintenance consumes over 70% of the total lifecycle cost of the software product. Documenting source code is therefore, imperative to software maintenance as it helps improve readability of the programs, enabling maintenance teams to clearly understand the working of the source program. It also helps them make relevant changes to keep the software compatible with evolving technologies and standards. Effective documentation, that is kept up to date as and when the code is refactored also makes the software easier to debug, ensures a smooth transition of the software between developer teams and facilitates software reuse.

Documentation is particularly essential for the functions in a program, as it aids programmers in grasping the role played by the function and such an understanding is crucial to comprehend the program's overall behaviour. Computer Aided Software Engineering or CASE tools such as Doxygen, Resharper and JavaDoc can only parse the existing documentation in source code files, and present them in a suitable format, but programmers still have to invest a significant amount of time and effort in composing the documentation. Hence, in many cases, documentation of source code is not performed even though it greatly improves the quality of the software product and is very beneficial for software development and maintenance.

Automating code documentation using machine learning-based systems based on big data input can solve the above issue. Most of the existing machine learning methods are based on neural machine translation which is a form of sequence to sequence learning. The words and special characters in a function present in the source code form the input sequence and the natural language documentation forms the output sequence. These methods use sequence models and are based on recurrent neural networks. Sequence models suffer from a range of problems such as vanishing gradients and increased computational complexity. Also, they are not able to capture long-range dependencies and are not interpretable, making it harder to discern their working.

In this work, we develop NeuralDoc, a novel end to end data driven system, for automating source code documentation by modelling it as a neural machine translation task. We use the Transformer architecture [2] for our machine translation model. Several research works have shown that Transformers perform well on many Natural Language Processing (NLP) tasks. In contrast to the existing models based on recurrent neural networks, the Transformer which leverages the self-attention and multi-headed attention mechanism can capture long-range dependencies. The Transformer also lends itself to parallelization as it does not process input data sequentially leading to significantly faster training and inference times. We also integrate the copy attention mechanism and incorporate the use of BERT [3], which is a technique for NLP pre-training into the basic Transformer architecture to create a novel approach for automating code documentation. We train our model on two datasets consisting of Java and Python source programs and their documentation, and carry out experiments to demonstrate the effectiveness of our proposed method. We also build an intuitive interface for users to interact with our models and deploy our system as a web application.

2 Literature Review

Early methods, such as the one by Haiduc et al. [4] and Moreno et al. [5], were non-neural approaches and were based on templates and heuristics, to automatically generate descriptions of source code. However, in recent times several systems based on neural networks using big data input have been proposed as they reduce human effort and offer greater flexibility.

Iyer et al. [6] proposed a machine learning model whose architecture is based on the Long Short-Term Memory (LSTM) unit which is a variant of the recurrent neural network that can easily capture long-range dependencies. The tokens in the input code snippet are transformed into vectors using an embedding matrix, which are fed into model. The natural language documentation is generated word by word, using the context of previously generated words which is obtained using the product of conditional probabilities, learned by the LSTM unit.

Allamanis et al. [7] described a method to extract a short phrase of summary from a snippet of code. Unlike most other approaches, they used convolutional layers combined with rectified linear activation units to extract features from the input data. They introduce an attentional neural network that employs convolutions on the input tokens to detect local time-invariant attention features and long-range dependencies. The final result was an implementation that can give a possible name for a code snippet constituting a function.

Shido et al. [8] formulated code documentation as a language translation task. They used abstract syntax trees along with the Tree-LSTM, a variant of LSTM modified to operate on tree structured data. Rather than converting the syntax trees into sequences as done in previous works, they designed a machine translation model which can directly learn the tree structure and can leverage the structural and syntactic information of the source code, resulting in a better quality documentation. Some of the more recent approaches such as the one by Wan et al. [9] incorporate an abstract syntax tree structure as well as sequential content of code snippets into a deep reinforcement learning framework. Wei et al. [10] propose a training framework where the two tasks of code generation and code summarization are performed simultaneously in order to boost the performance of both the tasks.

3 NeuralDoc

In this work we develop an end to end data driven, neural network-based system called NeuralDoc, for automatically generating source code documentation. We describe our system in detail in the following sections. Section 3.1 outlines our approach, giving a high level overview of our system. Section 3.2 describes our method in detail, explaining each component of our machine learning model architecture and Sect. 3.3 provides an algorithmic model of its working.

3.1 Approach

Figure 1 shows the high level block diagram of our system, NeuralDoc. We model

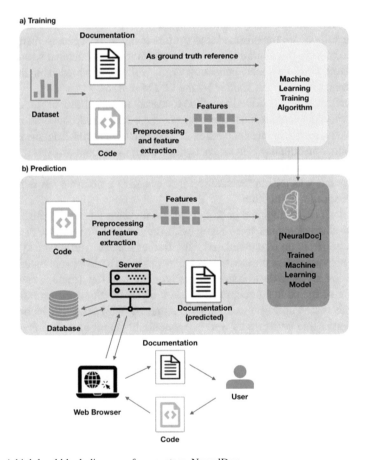

Fig. 1 A high level block diagram of our system, NeuralDoc

the problem of automatically generating source code documentation as a language translation task. In our system NeuralDoc, we use the Transformer architecture [2] to design the machine learning model which will perform Neural Machine Translation (NMT). The input code decomposed into tokens forms the input sequence while the output sequence is constituted by the corresponding documentation.

The first phase consists of selecting a dataset consisting of code and reference ground truth documentation pairs, pre-processing the dataset, and preforming feature engineering and feature extraction. The data is then used by a supervised machine learning algorithm to train a model, which in our case is the Transformer. The ability of the Transformer to capture long-range dependencies is particularly suitable for source code as the length of code is uncertain. A code snippet can be hundreds or even thousands of words long, and a code element may depend on another element many tokens away. The performance of various models obtained by tuning the base model architecture and various hyper-parameters is evaluated using suitable metrics and the best model is selected.

The trained model is placed on a server which is also connected to a database. The end user interacts with the server via a web browser. Our system NeuralDoc will be accessible to the users via the internet as a website on which they can upload their source code snippet. This code is sent to the server which feeds it as input to the trained machine learning model and performs inference to obtain an output documentation. This predicted documentation is sent back to user's web browser and is displayed on the web page. The web application as a whole, abstracts the working of the machine learning model, provides an intuitive interface and enhances user experience.

3.2 Method

We use the Transformer [2], which leverages the attention mechanism as the basic architecture for our machine learning model. We incorporate the copy attention mechanism along with the Bidirectional Encoder Representations from Transformers (BERT) [3] model into the basic Transformer to improve the quality of automatically generated documentation.

3.2.1 Attention

The attention mechanism receives three vectors as its input, namely the query, key and value. It performs a dot product of the query and keys and applies a softmax function over the result of the dot product to get attention scores. It then performs a sum of the values, weighing each of them using the attention scores to obtain the final output vector. The query and the keys determine which values to focus and can hence be described as 'attending' to the values. Let attn (q, K, V) represent the attention layer, where q, K and V indicate query, key and value, respectively. Here q

is a d_q-dimensional vector ($d \in Z$), K and V are two sets with $|K| = |V|$. Each $k_i \in K$ and $v_i \in V$ are d_k / d_v-dimensional vectors, and $i \in [|K|]$. The attention model works as follows:

$$attn(q, K, V) = \sum_{i=1}^{|v|} \alpha_i W_v v_i, \alpha_i = \frac{\exp\left((W_q q)^T (W_k K_i)\right)}{Z}, Z = \sum_{i=1}^{|K|} \exp\left((W_q q)^T (W_k K_i)\right)$$

where W_q, W_k and W_v are the parameters to be learned. As in the Transformer architecture proposed by [2] we implement attn as a multi-head attention model. In our experiments, we set $d_q = 64$, $d_k = 64$ and $d_v = 64$.

3.2.2 Transformer

The Transformer model [2] architecture consists of an encoder and decoder. The encoder consists of N identical layers. Each layer in turn has two blocks within it, the attention block and the feed forward neural network block. All blocks within the layer always output vectors of the same size, d_{model}, to ensure compatibility. In our experiments, we set $N = 6$, $h = 8$ and $d_{model} = 512$ for the encoder. The decoder also consists of N identical layers. Each layer in turn has three blocks. Two of these are same as the ones present in the encoder, and the third block computes the attention function using the output of the encoder using h attention heads. In our experiments, we set $N = 6$, $h = 8$ for the decoder.

3.2.3 Copy Attention

We incorporate the copy attention mechanism [11] in the Transformer architecture to enable the Transformer to not only generate new words that belong to the vocabulary but also copy word tokens from the input source code. This is done by using an extra attention layer to learn the copy distribution on top of the decoder stack. The copy attention allows the Transformer to copy rare tokens such as function and variable names from the source code and thus improves the quality of the generated documentation significantly.

3.2.4 Integrating the BERT Model

BERT [3] is a recently proposed technique that is used for NLP pre-training. It uses a large amount of easily available unlabelled text, and a deep neural network consisting of bidirectional sequence models to obtain an embedding for input text tokens. Using the pre-trained BERT models and training just a few additional layers on top it leads to superior results on a range of NLP tasks.

Inspired by [12], we obtain the representation of the input code sequence from a pre-trained BERT model and feed it into all layers in the Transformer architecture, rather than using it as an input embedding only. We introduce two new additional blocks to perform the attention mechanism between the BERT model's output embedding and the Transformer encoder and between the BERT model's output embedding and the Transformer decoder. Using the attention function our system, NeuralDoc can effectively learn how to merge the BERT model's embedding with the hidden layer representations in the Transformer encoder and decoder. In our experiments, we use the pre-trained $BERT_{basic}$ model with 12 layers and hidden dimension 768, provided by PyTorch.

3.3 Algorithm

Figure 2 shows a diagrammatic representation of our algorithm. It shows a pre-trained BERT model, the Transformer encoder and the Transformer decoder. The number of layers in the encoder and decoder is L, which is a hyper-parameter. The attention blocks compute the attention function as described in Sect. 3.2.1.

The dotted lines shows the residual connections between the layers, and the layer normalisation blocks are also shown in the figure. Let C and D represent the set of source programs and corresponding documentation, respectively. Both these sets contain sequences of text tokens. For any sequence of code tokens, $c \in C$ and any sequence of tokens constituting the documentation $d \in D$, let l_c and l_d be the number of tokens in c and d. The ith token in c / d is denoted as c_i / d_i. The outputs from the last layers of the pre-trained BERT model and encoder are H_B and $H_E{}^L$, respectively. Figure 3 shows the algorithm depicting the conversion of input source code tokens

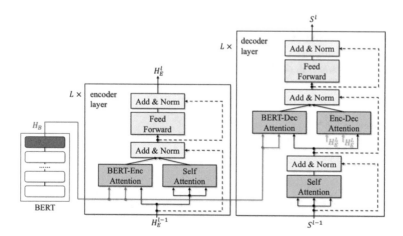

Fig. 2 Architecture of our Transformer-based deep-learning model adapted from [12]

Step-1: Given any input source code sequence, $c \in C$, BERT first encodes it into representation $H_B = \text{BERT}(c)$. H_B is the output of the last layer in BERT. The $h_{B,i} \in H_B$ is the representation of the i-th token in c.

Step-2: Let H_E^l denote the hidden representation of l-th layer in the encoder, and let H_E^0 denote word embedding of the input source code sequence c. Denote the i-th element in H_E^l as h_i^l for any $i \in [l_c]$. In the l-th layer, $l \in [L]$,

$$\tilde{h}_i^l = \frac{1}{2}\Big(\text{attn}_S(h_i^{l-1}, H_E^{l-1}, H_E^{l-1}) + \text{attn}_B(h_i^{l-1}, H_B, H_B)\Big), \forall i \in [l_x],$$

where attn_s and attn_a are attention models with different parameters. Then each $\tilde{h}_i^l$ is further processed by FFN and we get the output of the l-th layer:

$$H_E^l = \Big(\text{FFN}(\tilde{h}_1^l), \cdots, \text{FFN}(\tilde{h}_{l_x}^l)\Big)$$

where following [2] FFN is defined as

$$\text{FFN}(x) = W_2 \max(W_1 x + b_1, 0) + b_2,$$

where x is the input, W_1, W_2, b_1, b_2 are the parameters to be learned and max is an element-wise operator. Layer normalization is also applied and the encoder will eventually output H_E^L from the last layer.

Step-3: Let $S_{<t}^l$ denote the hidden state of l-th layer in the decoder preceding timestep t, i.e., $S_{<t}^l = (s_1^l, \ldots, s_{t-1}^l)$. Note s_1^0 is a special token indicating the start of a sequence, and s_t^0 is the embedding of the predicted word at time-step $t - 1$. At the l-th layer, we have

$$\hat{s}_t^l = \text{attn}_S(s_t^{l-1}, S_{<t+1}^{l-1}, S_{<t+1}^{l-1});$$

$$\tilde{s}_t^l = \frac{1}{2}\Big(\text{attn}_B(\hat{s}_t^l, H_B, H_B) + \text{attn}_E(\hat{s}_t^l, H_E^L, H_E^L)\Big), \quad s_t^l = \text{FFN}(\tilde{s}_t^l).$$

Fig. 3 Algorithm depicting the conversion of input source code sequence to documentation

into corresponding documentation.

The above equation computed in all the layers is mapped via a linear transformation and softmax to get the tth predicted word of the output documentation for the input code sequence. The decoding process continues over several time steps until a special token called the end-of-sentence token is produced by the decoder. Here, attn_S is the self-attention function, attn_B is the attention function using the BERT representations and attn_E is the attention function using the output of the encoder.

4 Experiment

4.1 Datasets

We conduct our experiment on a Java dataset and a Python dataset from [9], containing < method, comment > pairs. The Java and Python datasets were created using source code obtained from GitHub, a popular open-source projects hosting platform. Also following [10], code tokens of the form *CamelCase* and *snake_case*

Table 1 Statistics of the experiment datasets

Dataset	Java	Python
Train	69,708	55,538
Validation	8,714	18,505
Test	8,714	18,502
Unique tokens in code	66,650	307,596
Unique tokens in summary	46,895	56,189
Average tokens in code	120.16	47.98
Average tokens in summary	17.73	9.48

are separated into sub-tokens as it significantly reduces the number of unique tokens, thereby decreasing the size of the vocabulary. The statistics of the dataset are shown in Table 1, along with the train, validation and test splits of the labelled examples.

4.2 Evaluation Metrics

We evaluate the performance of the machine learning model using three metrics, namely BLEU, METEOR and ROUGE-L. BLEU is a metric that correlates highly with human judgements. It calculates n-gram precision between generated and target sentences. A perfect match results in a score of 1.0, whereas a perfect mismatch results in a score of 0.0. METEOR calculates a score which is a combination of unigram precision and unigram recall, and can evaluate how well ordered a generated sentence is. ROUGE-L is based on the length of the longest common sequence and computes recall scores between the generated and the ground truth documentation text. It takes into account sentence level structure similarity naturally and identifies the longest co-occurring sequences.

4.3 Hyper-Parameters

The transformer encoder and decoder consist of 6 layers, with a 512 dimensional hidden representation. The attention layers in the encoder and decoder use 8 attention heads and the query, key and value vectors have a dimensionality of 64. We use the pre-trained $BERT_{basic}$ model with 12 layers and hidden dimension 768. We use a mini-batch size of 32, and a dropout rate of 0.2. The Adaptive Moment estimation or Adam algorithm is used with the learning rate set to 0.0001 for optimisation. We train our models for 200 epochs, and validate the model performance after each epoch. We stop the training process if the BLEU score of the model on the validation set does not increase for more that 20 successive epochs. During inference, we perform decoding using two different methods, one using beam search with beam size set to 1 or greedy decoding and the other using beam search with beam size equal to 4.

4.4 Implementation

The experiments are performed with Python 3.6 and CUDA 10.1, with the help of a Google Colaboratory notebook. The Colaboratory notebook runs in a cloud VM, with a 2.2 GHz Intel Xeon CPU, with 13 GB RAM, on top of the Ubuntu 18.04.3 LTS operating system. We utilise one NVIDIA Tesla T4 GPU, with 12 GB RAM for training our models. We use PyTorch 1.6, a popular, open-source machine learning framework to aid our implementation of the machine learning models as well as for obtaining the weights for the pre-trained BERT [3] model. We use Streamlit, an open-source application framework specifically designed for machine learning and data science applications to build an intuitive user interface and deploy our system as a web application.

5 Results and Analysis

We summarise the results obtained from our experiments in Table 2 and Table 3. We compare our BERT integrated, Transformer-based approach, NeuralDoc with seven other baseline methods. We show the BLEU, METEOR and ROUGE-L scores obtained for the test set in the Java and Python datasets. From the evaluation results in Table 2, we can see that our method NeuralDoc performs better than all the baselines on the Java dataset in terms of all three metrics and performs on par with the best methods on the Python dataset. The relatively lower scores on the Python dataset

Table 2 Quantitative results of our approach, NeuralDoc compared with other methods

Method	Java			Python		
	BLEU	METEOR	ROUGE-L	BLEU	METEOR	ROUGE-L
CODE-NN [6]	27.60	12.61	41.10	17.36	09.29	37.81
Tree2Seq [13]	37.88	22.55	51.50	20.07	08.96	35.64
Hybrid2Seq [9]	38.22	22.75	51.91	19.28	09.75	39.34
DeepCom [15]	39.75	23.06	52.67	20.78	09.98	37.35
API + CODE [15]	41.31	23.73	52.25	15.36	08.57	33.65
Dual Model [10]	42.39	25.77	53.61	21.80	11.14	39.45
C2NL [14]	44.58	26.43	54.76	32.52	19.77	46.73
NeuralDoc (ours)	44.96	26.90	54.98	32.36	19.64	46.56

Table 3 Quantitative results of our approach, NeuralDoc with different decoding methods

Method	Java			Python		
	BLEU	METEOR	ROUGE-L	BLEU	METEOR	ROUGE-L
NeuralDoc (with beam size = 1) (greedy decoding)	44.66	27.06	54.96	32.00	19.88	46.37
NeuralDoc (with beam size = 4)	44.96	26.90	54.98	32.36	19.64	46.56

are mostly due to the low average length of the code snippets and ground truth documentation as compared to the Java dataset.

From the evaluation results in Table 2, we can see that our method NeuralDoc performs better than all the baselines on the Java dataset in terms of all three metrics and performs on par with the best methods on the Python dataset. The relatively lower scores on the Python dataset are mostly due to the low average length of the code snippets and ground truth documentation as compared to the Java dataset.

The high BLEU score is an indication of a good level of similarity between the automatically generated and ground truth reference documentation. The high METEOR score indicates that the generated documentation is well ordered, and the high ROUGE-L score signifies that the length of the longest common subsequence between the generated and ground truth documentation is relatively high on average.

These positive results may be attributed to the use of the copy attention mechanism and integration of the pre-trained BERT model with the Transformer. Using the copy attention mechanism, the Transformer can not only generate new words that belong to the vocabulary but also copy word tokens from the input source code. It allows the Transformer to copy function names and variable names directly from the source program and thus improves the quality of the generated documentation significantly.

With the integration of BERT into the basic Transformer architecture, the output features of the pre-trained BERT model are fed into all layers of the Transformer encoder and decoder, ensuring that the well-pre-trained features are effectively utilized.

Using the attention function to bridge the pre-trained BERT model and the Transformer, our system, NeuralDoc can successfully learn how to merge the BERT model's embedding with the hidden layer representations in the Transformer encoder and decoder to improve the quality of the generated documentation. We also show the results are obtained by two different decoding methods, namely decoding with beam size 1 also known as greedy decoding and decoding with beam size 4, in Table 3.

We can hence infer that our model is effectively capturing the structural information in the source code snippet. In contrast to natural language, the syntactic and semantic structure of the source code carries a great amount of information that is useful for the neural machine translation task and our approach NeuralDoc is able to effectively leverage it to generate better quality, concise documentation. We provide qualitative results of our approach, NeuralDoc on the Java and Python datasets.

Example 1

```
private int clampMag(int value, int absMin, int absMax)
{
    final int absValue = Math.abs(value);
    if(absValue < absMin)
        return _NUM;
    if(absValue > absMax)
        return value > _NUM ? absMax : - absMax ;
    return value ;
}
```

"NeuralDoc prediction": ["clamp the magnitude of value for absmin and absmax . if the value is below the minimum , it will be clamped to zero . if the value is above the maximum , it will be clamped to the maximum ."]
"Human written reference": ["clamp the magnitude of value for absmin and absmax . if the value is below the minimum , it will be clamped to zero . if the value is above the maximum , it will be clamped to the maximum ."]
"bleu": 1.0
"rouge_l": 1.0

Example 2

```
public boolean isSatisfiedBy(Date date)
{
    Calendar testDateCal = Calendar.getInstance(getTimeZone());
    testDateCal.setTime(date);
    testDateCal.set(Calendar.MILLISECOND, _NUM);
    Date originalDate = testDateCal.getTime();
    testDateCal.add(Calendar.SECOND, - _NUM);
    Date timeAfter = getTimeAfter(testDateCal.getTime());
    return (timeAfter != null)&&(timeAfter.equals(originalDate));
}
```

"NeuralDoc prediction": ["indicates whether the given date satisfies the cron expression . note that milliseconds are ignored , so two dates falling on different milliseconds of the same second will always have the same result here ."]
"Human written reference": ["indicates whether the given date satisfies the cron expression . note that milliseconds are ignored , so two dates falling on different milliseconds of the same second will always have the same result here ."]
"bleu": 1.0
"rouge_l": 1.0

Example 3

```
def ntohl(bs):
    return struct.unpack('I', bs)[0]
```

"NeuralDoc prediction": ["convert integer in n from network-byte order to host-byte order ."]
"Human written reference": ["convert integer in n from network-byte order to host-byte order ."]
"bleu": 1.0
"rouge_l": 1.0

Example 4

```
def make_password(password, salt=None, hasher='default'):
    if password is None:
        return UNUSABLE_PASSWORD_PREFIX +
get_random_string(UNUSABLE_PASSWORD_SUFFIX_LENGTH)
    hasher = get_hasher(hasher)

    if not salt:
        salt = hasher.salt()

    return hasher.encode(password, salt)

"NeuralDoc prediction": ["turn a plain-text password into a hash
for database storage same as encode() but generates a new random
salt ."]
"Human written reference": ["turn a plain-text password into a
hash for database storage same as encode() but generates a new
random salt ."]
"bleu": 1.0
"rouge_l": 1.0
```

6 Conclusion

In this work we propose NeuralDoc, an automated system to generate documentation from source code using machine learning techniques. We formulate the task as a case of language translation. We design a novel approach using a Transformer architecture, integrated with the copy attention mechanism and the use of the pre-trained BERT model. We perform experiments to demonstrate that our method outperforms the state-of-the-art approaches on the Java dataset. The adaptation of our approach to shorter length sequences, the compression of our BERT integrated Transformer models into lighter versions and the incorporation of abstract syntax tree representations of source code into Transformer-based approaches for automatic documentation would be interesting directions for future work.

References

1. X. Xia, L. Bao, D. Lo, Z. Xing, A.E. Hassan, S. Li,. Measuring program comprehension: a large-scale field study with professionals. IEEE Trans. Softw. Eng. (2017)
2. A. Vaswani, N. Shazeer, N. Parmar, J. Uszkoreit, L. Jones, A. Gomez, L. Kaiser, I. Polosukhin, Attention Is All You Need, in *NeurIPS* (2017)
3. J. Devlin, M.-W. Chang, K. Lee, K. Toutanova, Bert: pre-training of deep bidirectional transformers for language understanding, in *NAACL* (2019)
4. S. Haiduc, J. Aponte, L. Moreno, A. Marcus, On the use of automated text summarization techniques for summarizing source code, in *WCRE* (2010)

5. L. Moreno, J. Aponte, G. Sridhara, A. Marcus, L.L. Pollock, K. Vijay- Shanker, Automatic generation of natural language summaries for java classes, in *ICPC* (2016)

6. S. Iyer, I. Konstas, A. Cheung, L. Zettlemoyer, Summarizing source code using a neural attention model, in *ACL* (2016)

7. M. Allamanis, H. Peng, C. Sutton, A convolutional attention network for extreme summarisation of source code, in *ICML* (2016)

8. Y. Shido, Y. Kobayashi, A. Yamamoto, A. Miyamoto, T. Matsumura, Automatic source code summarisation with extended tree-LSTM, in IJCNN (2019)

9. Y. Wan, Z. Zhao, M. Yang, G. Xu, H. Ying, J. Wu, P.S Yu, Improving automatic source code summarization via deep reinforcement learning, in *Proceedings of the 33rd ACM/IEEE International Conference on Automated Software Engineering* (2018)

10. B. Wei, G. Li, X. Xia, Z. Fu, Z. Jin, Code generation as a dual task of code summarization, in *NeurIPS* (2016)

11. A. See, P.J. Liu, C.D. Manning, Get to the point: summarization with pointer-generator networks, in *ACL* (2017)

12. J. Zhu, Y. Xia, L. Wu, D. He, T. Qin, W. Zhou, H. Li, T. Liu, Incorporating BERT into neural machine translation, in *ICLR* (2020)

13. A. Eriguchi, K. Hashimoto, Y. Tsuruoka, Tree-to-sequence attentional neural machine translation, in *ACL* (2016)

14. W.U. Ahmad, S. Chakraborty, B. Ray, K.-W. Chang, A transformer-based approach for source code summarization, in *ACL* (2020)

15. X. Hu, G. Li, X. Xia, D. Lo, Z. Jin, Deep code comment generation, in *ICPC* (2018)

Transfer Fault Detection in Finite State Machines Using Deep Neural Networks

Habibur Rahaman, Santanu Chattopadhyay, and Indranil Sengupta

Abstract This work presents a testing scheme for Finite State Machines (FSM) based on Deep Neural Network (DNN). This technique determines whether a given implementation FSM-B is equivalent to its specification FSM-A. The input/output sequences (I/O pairs) for a given FSM train the proposed DNN. First, I/O pairs of FSM-A are generated using an adaptive distinguishing algorithm, and most of these sequences (around 80%) are used for training the DNN. After training, the remaining 20% I/O pairs are used for validating the derived DNN. After training and validation, the correctness of *FSM-B* is checked. A small number of vectors is applied to FSM-B and the generated outputs are compared with the DNN predicted outputs. Based on the similarity percentage between them, FSM-*B* is declared either as correct or faulty implementation of FSM-A. To check the effectiveness of the scheme, transfer-type faults are injected to construct mutant FSMs. The results of experimentation performed on the MCNC FSM benchmark prove the efficacy of this scheme. As only a subset of tests is needed to check the presence of fault if any, the testing time is remarkably less—resulting in an average reduction of 87.68% compared to the conventional technique. To the best of our knowledge, this DNN based testing scheme is being presented for the first time.

Keywords Fault detection · Conformance testing · DNN · FSM · Black box · Machine verification

H. Rahaman (✉) · S. Chattopadhyay · I. Sengupta
Indian Institute of Technology Kharagpur, Kharagpur 721302, India
e-mail: habibur@iitkgp.ac.in

S. Chattopadhyay
e-mail: santanu@ece.iitkgp.ac.in

I. Sengupta
e-mail: isg@iitkgp.ac.in

© The Author(s), under exclusive license to Springer Nature Singapore Pte Ltd. 2022   139
B. Mishra et al. (eds.), *Artificial Intelligence Driven Circuits and Systems*,
Lecture Notes in Electrical Engineering 811,
https://doi.org/10.1007/978-981-16-6940-8_12

1 Introduction

FSMs lie at the heart of various complex computing systems like sequential circuits, telecommunications systems, communications protocols, embedded systems and many other related fields. The control portions of today's communication protocols [1–3] are mostly modelled by FSM. These complex systems are also less reliable. Hence, testing such FSM is indispensable to ensure their correct functioning. An FSM has a finite set of states and produces output on state transitions upon receiving the input. In testing problems, an FSM consists of a transition diagram, however, its present state is not known. An input sequence known as *homing sequence* is applied to the FSM so that from its input–output behaviour, its state information can be derived. In the state identification problem, actually, the initial state of the FSM would be identified. A test sequence known as a *distinguishing sequence* solves this problem. The unique input–output (UIO) sequence verifies the specified state of the FSM in the state verification problem.

Another testing method of the FSM, known as 'fault detection' or 'machine verification', is formally known as conformance testing or 'test generation'. Here, there is a specification FSM-A, and an implementation FSM-B (also known as "black box") with only its input–output behaviour being observable. A test sequence known *as checking sequence* is generated and is applied to FSM-B and by observing its output, it is confirmed whether FSM-B is a good implementation of FSM-A or not. Here, the problem of conformance testing based on DNN has been proposed in which adaptive distinguishing sequence (ADS) has been utilized.

Various works on testing and diagnosis of FSM-based computer systems have been reported earlier. The problem of fault detection in FSM reported by Moore's work on 'gedanken-experiments' [4] in 1956 is the harder problem of machine identification. The testing problems of FSM, based on mainly automata theory and switching circuits, have been published in [2]. Some important techniques on FSM testing are the distinguishing sequence-based D-method [5], the UIO sequence-based U-method [6], the characterization sets-based W-method [7], and the transition tours-based T-method [8].

Recently, some metaheuristic search-based techniques have been introduced for test data generation. The search-based functional testing techniques having the state machine specifications have been discussed in [9, 10]. The test generation in the state machine using a genetic algorithm has been presented in [9], whereas three search techniques (genetic algorithms, simulated annealing and particle swarm optimization) have been utilized in [10].

The technique proposed in [11] uses an artificial neural network in the form of an automated oracle for testing software systems. A state count-based FSM conformance testing has been discussed in [12]. Test case generation scheme in FSM based on graph algorithm has been reported in [13, 14].

Distinguishing sequence (DS) has been used in [15–17] to generate the test sequence for FSMs. The DS followed by a transfer sequence is applied to the input of implementation FSM and a transition from state q_i to q_j under input x is verified by checking the output produced on input x to check an output fault or by knowing the state reached upon receiving input x at state q_j of implementation FSM to identify a transfer fault. The minimization of the length of checking sequences produced from FSM for testing has been discussed in [16].

A testing and model checking technique based on genetic algorithm has been presented in [18]. The 'gedanken' experiment-based testing of FSM part of the protocol is explained in [19].

A test generation scheme for testing FSM in sequential circuits is discussed in [20]. An identification method has been proposed in [21] for protocol testing. Using a probabilistic approach, a conformance testing of protocols is presented in [22]. Using homing and synchronizing sequence, a test generation scheme for sequential circuits is presented in [23]. An FSM conformance testing scheme for software system has been described in [24]. An artificial neural network-based FSM testing for software system has been presented in [25].

Here, a DNN guided testing scheme is developed to verify whether the implementation FSM-B is faulty or fault-free. First, the input sequence for the specification FSM-A is produced from the transition diagram by *Adaptive Distinguishing* sequence algorithm [2, 15] and the respective output sequence is noted. The constructed I/O pairs are used in DNN. While the majority (~80%) of the pairs are used in the training the DNN, 20% of pairs are used for validation. Once the DNN is validated, it is applied to verify the correctness of FSM-B. FSM-B is supposed to be in its known state before the test is conducted. If it is not in a known state, a *homing sequence* is applied to FSM-B to move it to a known state. The same input sequences are fed to the implementation FSM and the trained DNN and the dissimilarities/similarities between the outputs of the DNN and FSM-B are observed. If these outputs match, FSM-B is declared as good, otherwise B is faulty.

To check the effectiveness of the scheme on transfer type faults, such faults are injected to obtain the Mutant FSM-B. In the presence of transfer faults, the implementation goes to some wrong state. A multi-layer neural network has been used to detect faults within the mutated version of the FSM. The test time is very less as only a small subset of test patterns need to be used.

The remaining part of the article is oriented as given below. Section 2 describes the preliminaries of the works, Sect. 3 explains the presented scheme. Section 4 highlights the results and discussions, and lastly, Sect. 5 presents the conclusion and the future work.

2   Preliminaries

2.1   *Finite State Machine (FSM)*

An FSM modelled by a mealy machine consists of a finite set of states and on receiving input produces output on the state transition. Formerly, it is specified by a six tuple $(P, O, Q, q_0, \delta, \lambda,)$, where

$P \rightarrow$ input alphabet,

$O \rightarrow$ output alphabet.

$Q \rightarrow$ a finite set of states.

$q_0 \in Q \rightarrow$ the initial state.

$\delta: Q \times P \rightarrow$ state transition function.

$\lambda: O \times P \rightarrow$ output function.

On receiving an input 'a' $\in P$, an FSM moves from current state $q \in Q$ to the next state given by δ (q, a) and produces output specified by λ (q, a). Usually, FSM is represented by a directed graph known as a state transition diagram, where its vertices represent states and edges specify state transitions and each edge is identified with input/output associated with the transition. FSM as shown in Fig. 1 is currently in q_0 state. When input b *is applied*, it arrives at state q_1 and generates '1' output.

Here, FSM-A is completely specified, deterministic, reduced and strongly connected. Every state in strongly connected FSM is reachable to other states via one or more transitions. There exists an input sequence for each pair of states (q_i, q_j) for which the FSM moves from state q_i to state q_j. From each state, there exists a transition for each input in a completely specified FSM. In minimal or reduced FSM, the number of states is less than or equal to the original FSM. The input sequence known as the checking sequence is applied to FSM-B to prove its equivalence. The checking sequence for FSM-A distinguishes the FSMs equivalent to FSM-A from other FSMs.

Fig. 1 State transition diagram

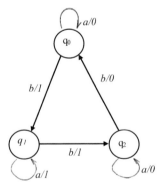

2.2 Homing Sequences

The input sequence which drives an FSM to be in the known state is known as a *homing sequence* [2–4]. Each minimal FSM has a homing sequence which will be generated in polynomial time. In the successor tree, a number of states indicate uncertainty. The set $\{q_0 \, q_1 \, q_2\}$ as shown in Fig. 1 is known as the initial uncertainty set. An input sequence generating current state singletons (containing only one element) as an uncertainty vector is a homing sequence.

In Fig. 1, the sequence '*ba*' is a homing sequence taking FSM from states q_0, q_1 and q_2, to q_1, q_2 and q_0, respectively, and the final state will be determined by the respective output sequences 11, 10 and 00, respectively.

2.3 W-method

W-set has been used in [15] for generating test sets in FSMs. Let $(P, O, Q, q_0, \delta, \lambda)$ be a minimal and complete FSM. A W-set for this FSM is a finite set of symbols such that for each pair of states (q_i, q_j), there is at least one symbol that distinguishes q_i from q_j. So, if q_i and q_j are two states, there exists an input '*a*' in *W* such that $\lambda(a, q_i) \neq \lambda(a, q_j)$, where '*a*' belongs to the input alphabet set *P*.

From Fig. 1, it is observed that *{(a), (b), (ba), (bb), (bba), (bbb)}* is W-set or known characterization set for FSM. So, when FSM-A starts from q_0 and in response to input sequence *bbb*, 110 will be generated as the output sequence. But if FSM starts from q_1 and on receiving input sequence *bbb*, 101 will be the generated as the output sequence. So, O (*bbb*, q_0) $\neq$ O (*bbb*, q_1) indicates sequence *bbb* as a DS which distinguishes q_0 and q_1.

2.4 Transfer Fault

Due to the presence of transfer faults, the implementation FSM makes wrong transitions. Figure 1 shows that for transition $q_1 \rightarrow q_1$, input '*a*' is required. However, due to the presence of transfer fault, the implementation FSM makes transition $q_1 \rightarrow q_1$, on input '*b*' as shown in Fig. 2. In this case, the FSM goes to q_2 state instead of q_1. This type of transfer fault has been considered in this paper.

2.5 Distinguishing Sequence (DS) Method

DS-based method has been used for test generation. Distinguishing sequence is of two types: preset distinguishing sequence (PDS) and adaptive distinguishing sequences

Fig. 2 State transition
diagram of a faulty FSM

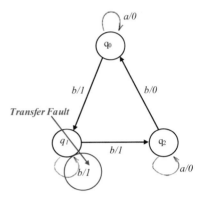

(ADS). Here, an ADS generation scheme has been used. An ADS is not really a
sequence but a decision tree. The length of the ADS is often shorter compared to that
of PDS. The shortest PDS of an FSM is of exponential length [15], while an ADS
has a length of $\{n(n-1)/2\}$, n being the number of states [15].

2.5.1 Preset Distinguishing Sequence (PDS)

PDS is an input sequence such that the output sequence generated by FSM on the
application of input 'a' is different for each initial state, i.e. $\lambda(q_i, a) \neq \lambda(q_j, a)$ for
each pair of states $q_i, q_j, i \neq j$.

In Fig. 1, 'ab' is a DS for the FSM, because $\lambda(q_1, ab) = 01$, $\lambda(q_2, ab) = 11$ and
$\lambda(q_3, ab) = 00$.

2.5.2 Adaptive Distinguishing Sequence (ADS)

ADS is a decision tree T with exactly n leaves, n being the number of states of the
FSM. Input symbols, output symbols and states of an FSM indicate internal nodes,
edges and leaves, respectively, such that (1) edges having distinct output symbols
emanate from a common node and (2) for every leaf of T, $z = \lambda(q_i; a)$ if the leaf is
specified by state q_i of the FSM and if a, z are input, output, respectively, formed by
the node and edge labels on the path from the root to the leaf. The sequence length
is equal to the depth of the tree.

In Table 1, initial uncertainty is $\{ABCD\}$. For $x = 0$, the uncertainty is $\{CD\}$
on $z = 0$ or $\{BC\}$ on $z = 1$. This process has been described as an adaptive tree
as shown in Fig. 3a. If the uncertainty is $\{CD\}$, input $x = 1$ distinguishes state C
from D. If the uncertainty is $\{BC\}$, then input $x = 0$ distinguishes state B from C. In
designing the adaptive tree, each path that leads to an uncertainty containing repeated
entries is terminated and the adaptive sequences are the paths that lead to singleton
uncertainties. In Fig. 3b, each of the paths emanating from the initial uncertainty

Table 1 .

PS	NS, Z $x = 0, \quad x = 1$
A	C, 0 A, 1
B	D, 0 C, 1
C	B, 1 D, 1
D	C, 1 A, 0

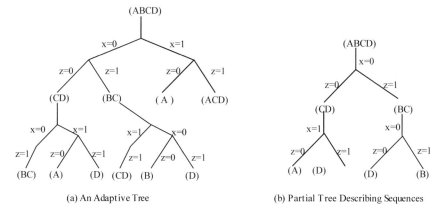

(a) An Adaptive Tree   (b) Partial Tree Describing Sequences

Fig. 3 Constructing adaptive distinguishing sequences for machine

{ABCD} leads to a singleton uncertainty. The adaptive distinguish sequences are 00 and 01.

2.6 Deep Neural Networks (DNNs)

Neural Network (NN) is a computational learning unit that tries to mimic the human brain, takes the input and processes it, and then finally finds the output. In the same way, the brain of humans work. A deep neural network (DNN) is a NN with more than one hidden layer.

A basic NN structure [26] is an input layer, hidden layers, output layer as shown in Fig. 4. In the NN, each input node is feeding the hidden layers, each hidden layer will be connected to other hidden layers, and in the last hidden layer, all nodes will be connected to the output (Fig. 4).

A basic structure of a single-neuron or node is shown in Fig. 5. A node is a computational unit of a neural network. Inside the node, initial random node weights are assigned. These weights are multiplied with input and summed up with some random bias after this computed value is passed to an activation function like reLu, sigmoid. We used the reLu activation function; finally, we will get of an output of

Fig. 4 Basic structure of a
neural network

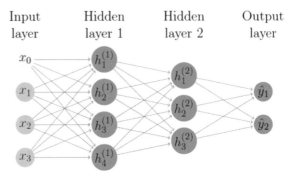

Fig. 5 Active node of neural
network

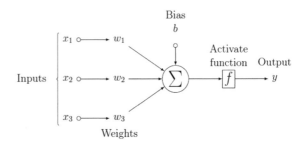

that particular node, similarly, it will compute all nodes output, and neural network
gets the train as process increases.

3 Proposed DNN Based Method

3.1 Proposed Algorithm and Its Explanation

The following steps have been used to develop a DNN guided strategy for transfer
fault detection in FSM.

Step 1: Derive input vector by Homing sequence to initialize given FSM.

Step 2: Generate the set of inputs for the FSM by ADS method. Apply input set to
the FSM to generate by the outputs using W-method. Now, derive the error terms
from the actual inputs and inputs generated from step-2. Train the DNN by error
terms such that it can predict the inputs given its respective outputs.

Step 3: Determine l_{max} (maximum bit length) of entire output vectors generated
(*step-2*). Check whether numbers of bit length of any vector is less than l_{max} or
equal. If it is less than l_{max} , make the bit length of that vector equal to l_{max} by
adding padding bits. Find respective actual inputs applying the outputs to DNN.
Also, make the number of bit length uniform for each input vector using padding

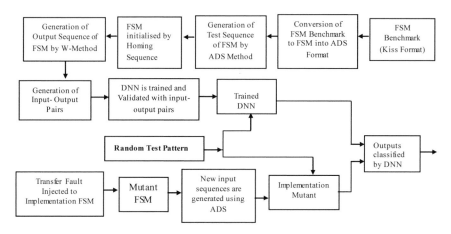

Fig. 6 Workflow of the proposed DNN-based testing method

bits. Now, derive the error terms from the actual inputs and inputs generated (step-2). Train the DNN by error terms such that it can predict the inputs given its respective outputs.

Step 4: Input fault is injected in the given FSM. Now, ADS method generates a new set of inputs with the help of this FSM having faulty inputs.

Step 5: Apply the output sequence (obtained in *Step 2*) to the trained DNN to generate the inputs. Now, compare the inputs obtained from the trained DNN and new inputs obtained from *Step 4*. The FSM-B has passed the verification for that output if they match.

Step 6: Continue with *step 5* for the generation of an adequate number of output sequences. Check the success rate whether it is greater than a predefined threshold or not and black-block FSM is declared as fault-free or faulty if success > threshold.

The flow of the DNN-based scheme is summarized in Fig. 6. First, the FSM benchmark (KISS-format) has been converted to the benchmark in ADS format and the resultant benchmark is passed through ADS to generate input sequences. The FSM is set to a known state using homing sequence. Then W-method produces output sequences. Next, the generated sequences have been applied to FSM to produce the respective output sequences using W-method.

80% of the output–input pairs obtained from ADS and W-method are applied to train the DNN. DNN has been validated by the remaining 20% pairs. As the bit length of the output/input vector is different, some additional bits known as padding bits are added to equalize their bit length. Table 2a shows the non-uniform bit length of input–output pairs of DK-17, and after adding padding bits, the uniform bit length has been shown in Table 2b. We consider the output–input pairs of DK-17 benchmark circuit. Here, '2' and '5' have been added as input padding bit and output padding bit, respectively. So after adding the padding bits, the output–input pairs will take the form as shown in Table 2b.

Table 2 Output–input sequences of *DK-17* benchmark

a) Normal		b) After Padding	
Output	**Input**	**Output**	**Input**
a c e e	*w yx w*	*a c e e 5 5 5*	*w yx w 2 2 2*
c f a f c f b	*y z x z y x w*	*c f a f c f b*	*y z x z y x w*
c b c	*y w z*	*c b c 5 5 5 5*	*y w z 2 2 2 2*

Now, transfer type fault is injected into the implementation FSM to construct the mutant FSM-B. This makes the good FSM faulty, referred to as mutant 'black box'. New test sequences are generated using ADS method corresponding to some previous output sequences. At the same time, with the same random output sequences, the predictions have been done. DNN predicted inputs are also recorded. Next, we compare the inputs obtained after applying ADS method to the faulty FSM (with input-faults) with those obtained from the DNN. The similarity percentage between black-box FSM and specification FSM (DNN) will decrease with the increase in the number of injected faults.

Lastly, using DNN, the outputs are classified to take the decision whether the black-block FSM is defective or not.

3.2   Deep Neural Network (DNN) Model

Tabular Learning model of 'fastai' library of Python [27] has been used. Tabular data source has been supported by Fastai library. This library is used for multi-bit prediction. Basically, 'fastai' is a deep learning library that provides low-level counterparts that can be matched and mixed to develop new methods.

Accuracy: We use the following accuracy measure in our experiments.

$$A = \frac{P \times 100}{C}.$$

where
   P = Total input sequences passing the test
   C = Total input sequences used in the experiment.

4   Experimental Results and Analysis

We have experimented using a system with 2.66 GHz Intel Core i7 CPU with 32 GB RAM and RTX 2060 6 GB GPU. We have taken eight FSM benchmark circuits as shown in Table 3. Here, we have shown how the accuracy of the 'black box' (the real FSM implementation) varies with the injected faults. Table 3 shows our experimental data for all the above-mentioned benchmark circuits. It indicates

Table 3 Similarity percentage of FSM-B with FSM-A with injected faulty input alphabets

Bench-mark	Input–Output	Number of States	Number of tests	Accuracy percentage of specification FSM-A	Similarity percentage of FSM-B for a variable number of faulty input alphabets	
					1	2
tbk	6/3	16	75,04,996	98.01	94	56
S298	3/6	135	6,44,529	96.5	92	30.92
DK-14	3/5	7	6,949	98	76	36
DK-17	2/3	8	927	90	60	16
TAV	4/4	4	16,112	80	76	62.5
MC	3/5	4	2,088	89.5	67.5	28
S27	4/1	5	18,964	84.39	62.4	53.7
Beecount	3/4	8	2457	85	67	49

the percentage accuracy of mutant FSM-B (the 'black box') for various numbers of faults. It shows the similarity difference between the output of the black-box FSM and that of DNN with the change of injected faults. We consider the cases in which one input alphabet (say a or b or …) is defective and, two (say, a and b, b and e, …,) input alphabets are defective shown in Table 3. Similarity % index of the mutant FSM for different number of defective input alphabets has been tabulated in the columns. Faults are injected at each input position containing a single input alphabet (say 'a' or 'b' or c etc.). Suppose, in an FSM, there are 10 positions containing input alphabet (say), 'a' and for 'a' as faulty, all 10 positions containing alphabet 'a' are replaced by another alphabet (say, t).

Table 3 shows the percentage similarity variation between output of trained DNN and output of 'black box' with the change in fault injection. Table 3 indicates that the accuracy percentage of faulty FSM-B is less than the accuracy percentage of the specification FSM-A. Overall consistent results for all other benchmarks prove the effectiveness of this DNN-based testing scheme.

Test time has been measured for mutant FSM-B with injected faults by W-method. DNN prediction time (test time) has also been recorded. In Table 4, the test time of various mutant benchmarks required by the DNN scheme with the test time required by the earlier W-method technique has been compared for the variable faulty input alphabets. Here, it is seen that the *DK-14* benchmark requires 15.86 times less time in DNN-based scheme compared to that in W-method. Again, the '*TAV*' benchmark requires 104 and 5.655 s in W-method and DNN-based scheme, respectively, and it results in a test time reduction of 94.58%. Figure 7 shows the plots of test time for mutant benchmarks using the W-method and the DNN-based method.

Table 4 Test time of mutant benchmarks needed by DNN guided method and W-method

Bench-mark	Test time (Sec.) needed by FSM-B IN W-method with a variable number of faulty input alphabets			Test time (Sec.) needed by FSM-B in DNN method with a variable number of faulty input alphabets			% Reduction in test time
	1	2	Average	1	2	Average	
tbk	11,099.25	11,100.40	11,099.83	227.60	227.30	227.45	97.95
S298	2958.00	2999.00	2978.5	517.70	526.0	521.85	82.48
DK-14	32.00	31.00	31.5	1.90	2.07	1.985	93.70
DK-17	13.00	12.00	12.5	2.67	2.50	2.585	79.32
TAV	103.70	104.80	104.25	5.31	6.00	5.655	94.58
MC	16.50	14.60	15.55	2.50	2.54	2.52	83.79
S27	108	103.80	105.9	3.23	3.25	3.24	96.94
Beecount	19.00	18.00	18.5	3.20	3.12	3.16	82.92
Average (%)							87.68

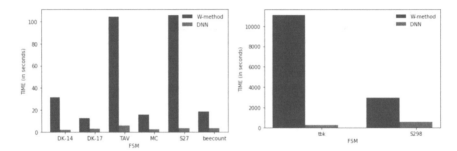

Fig. 7 Test time in FSM benchmarks

5 Conclusions

This work has presented a DNN guided testing scheme to test whether an implementation FSM is fault-free or not. This is the first-ever DNN-based FSM testing technique reported in the literature to the best of our knowledge. Once the trained DNN is obtained, only a small set of tests is needed to check the abnormality of FSM, if any. A small amount of test vectors requirement reduces the test time remarkably resulting in a reduction of 87.68% compared to the existing methods. Thus, the method has the potential of testing large FSM in polynomial time.

The results and analysis show the effectiveness of the DNN guided FSM testing scheme. In future, various other fault models may be incorporated into this model.

References

1. A.D. Friedman, P.R. Menon, *Fault Detection in Digital Circuits* (Prentice-Hall, Englewood Cliffs, NJ, 1971)
2. Z. K'ohavi, Switching and Finite Automata Theory, 2nd ed. (New York McGraw-Hill, 1978)
3. G.J. Holzmann, *Design and Itdidation of Protocols* (Prentice-Hall, Engle-Wood Cliffs, NJ, 1990)
4. E.F. Moore, Gedanken-experiments on sequential machines: automata studies, vol. 34 (Princeton University Press, Annals of Mathematics Studies, Princeton, NJ, 1956), pp. 129–153
5. F.C. Hennie, Fault detecting experiments for sequential circuits, in *5th Annals Symposium Switching Circuit Theory and Logical Design* (1964), pp. 95–110
6. K.K. Sabnani, A.T. Dahbura, A protocol test generation procedure. Comput. Netw. ISDN Syst. **15**(4), 285–297 (1988)
7. T.S. Chow, Testing software design modeled by finite-state machines. IEEE Trans. Softw. Engg. **SE-4**(3), 178–187 (1978)
8. S. Naito, M. Tsunoyama, Fault detection for sequential machines by transitions tours, in *IEEE Fault Tolerant Computing Symposium, IEEE Computer Society Press* (1981), pp. 238–243
9. R. Lefticaru, F. Ipate, Automatic state-based test generation using genetic algorithms, in *IEEE Intl. Symposium on Symbolic and Numeric Algorithms for Scientific Computing* (2008), pp. 188–195
10. R. Lefticaru, F. Ipate, Functional search-based testing from state machines, in *IEEE International Conference on Software Testing, Verification, and Validation* (2008), pp. 525–528
11. M. Vanmali, M. Last, A. Kandel, Using a neural network in the software testing process. Int. J. Intell. Syst. **17**(1), 45–62 (2002)
12. E. Vinarskii, A. Laputenko, J. López, N. Kushik, Testing digital circuits: studying the increment of the number of states and estimating the fault coverage. in *2018 19th International Conference on Micro/Nanotechnologies and Electron Devices (EDM)*
13. M.M. Mariano, É.F. Souza, N. Vijaykumar, Analyzing graph-based algorithms employed to generate test cases from finite state machines, in *2019 IEEE Latin American Test Symposium (LATS)*, pp. 1–6. https://doi.org/10.1109/LATW.2019.8704603
14. M.M. Mariano, E.F. de Souza, A.T. Endo, N.L. Vijaykumar, Comparing graph-based algorithms to generate test cases from finite state machines. J. Electron. Test. **35**, 867–885 (2019)
15. D. Lee, M. Yannakakis, Principles and method of testing finite state machines—survey. IEEE Proc. **84**(8) (1996)
16. R.M. Hierons, H. Ural, Reduced length checkingsequences. IEEE Trans. Comput. **51**(9), 1111–1117 (2002)
17. R.M. Hierons, H. Ural, Optimizing the length of checking sequences. IEEE Trans Comput. **55**(5), 618–629 (2006)
18. F. Tsarev, K. Egorov, Finite state machine induction using genetic algorithm based on testing and model checking. ACM GECCO'11, July 12–16, 2011, pp. 759–762
19. N. Kushik, J. López, A. Cavalli, N. Yevtushenko, Improving protocol passive testing through 'Gedanken' experiments with finite state machines, in *2016 IEEE International Conference on Software Quality, Reliability and Security*, pp. 316–322
20. G. Buonanno, F. Fummi, D. Sciuto, F. Lombardi, FsmTest: functional test generation for sequential circuits. INTEGRATION VLSI J. **20**, 303–325 (1996)
21. M. Yannakakis, D. Lee, Testing state machines: fault detection. J. Comput. Syst. Sci. (50) (1995) 209–227
22. S.H. Low, Probabilistic conformance testing of protocols with unobservable transitions, in *1993 International Conference on Network Protocols (ICNP 1993-IEEE Computer Society 1993)*, pp. 368–375
23. I. Pomeranz, S.M. Reddy, Application of homing sequences to synchronous sequential circuit testing. IEEE Trans. Comput. **43**(5) 569–580 (1994)
24. C. Baron, J.-C. Geffroy, Identification methods and conformance testing, in *IEEE 3rd Annual Atlantic Test Workshop (1994)*, (IEEE, 2002), pp. B1–B10

25. R. Zhao, S. Lv, Neural-network based test cases generation using genetic algorithm, in *13th IEEE International Symposium on Pacific Rim Dependable Computing 2007*, pp. 97–100
26. S. Haykin, Neural Networks, A Comprehensive Foundation (1999)
27. J. Howard, S. Gugger, Fastai: a layered API for deep learning. Information, **11**(2), 108 (2020)

Detecting Anomalies in Power Consumption of an Internet of Things Network Using Statistical Techniques

Edwin Jose, Ajai John Chemmanam, Bijoy A. Jose, and Asif Mooppan

Abstract Power consumption of household and commercial establishments is increasing every year. The vast amount of data generated by the smart energy metres cannot be monitored manually. Hence, automated power consumption monitoring and anomalous usage triggering are the need of the hour. Effective models for such processes are hindered due to the limited availability of labelled datasets. The lack of labelled datasets and high variance of data from various sources makes it difficult to develop a universal monitoring algorithm. Our study was accelerated by the acquisition of energy metre data of ATMs, through industrial collaboration. Analysis and study of real-world data from ATMs, lead to the development of custom labelled datasets. Artificial anomalies of various amplitudes were added to fixed locations on the data modelled. Custom labelled dataset generated was used to tune and evaluate the statistical anomaly detection algorithms and packages. The hyper parameters of the algorithms are tuned using hyperopt bayesian optimization. The results from the evaluations are tabulated and concluded that the hyper-tuned seasonalAD algorithm from ADTK package performed better amongst the statistical anomaly detection algorithms.

This project is funded by the DST ICPS Project entitled: "Energy Efficient Cyber Security Implementation of Internet Of Things". The power consumption anomaly detection is in collaboration with Vuelogix Pvt. Ltd.

E. Jose (✉) · A. J. Chemmanam
Department of Electronics, Cyber Physical Systems Lab, Cochin University of Science and Technology, Kochi, Kerala, India
e-mail: edwin.jose@cusat.ac.in

A. J. Chemmanam
e-mail: ajaichemmanam@cusat.ac.in

B. A. Jose
Department of Computer Science, Cochin University of Science and Technology, Kochi, Kerala, India
e-mail: bijoyjose@cusat.ac.in

A. Mooppan
Vuelogix Technologies Pvt Ltd, Kochi, Kerala, India
e-mail: asif@vuelogix.com

Keywords Power consumption · Anomaly detection · Time series decomposition · Statistical analysis · Real-time anomaly detection · ADTK · Prophet · Real-world data analysis · Custom dataset generation

1 Introduction

Large amount of data is generated from the Internet of Things (IoT) sensors as a part of Industry 4.0. Real-time data is generated from power metres, water metres, HVAC, etc. Power consumption data is one of such prominent data feeds that is being generated by IoT devices and monitored by smart energy metres. The data generated is so immense that humans cannot monitor manually [16]. There are different automation techniques available in the literature for gaining insight into power consumption. Two aspects that require special attention is power surges and power outages. We were posed with an IoT smart metre network, measuring power consumption in a banking institution, where the anomalous power usage could show a cybersecurity threat. Monitoring of ATMs, banking facility, etc. could be under direct threat in a power outage, while power usage during a holiday could indicate a robbery attempt. In this context, detection of anomaly in power consumption from IoT network gains importance.

Our research involved study of energy data from two hundred plus sources for a time period of six months. The sensors record periodical changes in energy consumption every 2 min. The data recorded are univariate time series in its properties. The initial study focused on using the statistical algorithms to clean and pre-process the data. Analysis of this refined real time data, supported in the modelling of the custom labelled dataset. In this paper, we present our observations on application of statistical anomaly detection techniques on this custom dataset.

1.1 Challenges and Difficulties

Anomaly detection algorithms face challenges in developing a effective classification /detecting model. The following are certain challenges that are specific to the domain of power consumption anomaly detection.

1.1.1 Lack of Anomaly Definition

The lack of standard definition for normal and anomalous data makes it difficult to define the available data. Furthermore, such definitions are dynamic considering the application domain. Thus defining anomaly and creating a dataset which is more aligned with the real-world data, will accelerate these studies [13].

1.1.2 Data Imbalance

Anomalous Events are rare events hence their datasets are sparse and imbalanced [10]. Thus machine learning predictions will be biased towards a high denser class and will not be able to classify or identify anomalies. Preprocessing algorithms can be applied to tackle such data biases. These prepossessing methods include:

(a) *Re-sampling*: This process replicates or reduces the counts of the data points such that all classes of data have equal density or weights. This process was achieved effectively using Synthetic Minority Over-sampling Technique (SMOTE) algorithm [6].
(b) *Synthetic power generation data*: This process creates data that is equivalent to the existing data using data models [20].

1.1.3 External Unpredictable Factors

Certain unpredictable factors like introducing a new device or various human activities shall bring dynamic variations in power consumption. All these dynamic variations may or may not be an anomaly. Thus such points cannot be effectively classified by the algorithms [10].

1.1.4 Unavailability of the Standard Dataset

Bench-marking the models without a standard dataset is difficult. This is because most researchers use various ground truth data of their preferences [22]. There are benchmark datasets for time series classification and general anomaly detection [17]. But univariate time series datasets specific to the domain of power consumption anomaly detection are limited.

1.1.5 Lack of Performance Assessment Methods

The real-world data are unlabelled time series in nature. Evaluation of anomaly detection models on such data will be difficult. To tackle this real world like labelled datasets has been modelled or manually labelled. Classical evaluation parameters like area under the curve (AUC), F_1 score can widely used for evaluating labelled datasets.

Our study attempted to address the above-mentioned challenges like: the lack of anomaly definition, data imbalance, lack of labelled dataset. The analysis of real-world dataset aided in the generation of custom labelled dataset. The custom labelled datasets along with some algorithms/ packages as mentioned above, helped to address the challenges.

This paper is organized as follows. Section 2 introduces on various definitions and background study on the types of anomalies, algorithms, and the evaluation parameters. Section 3 explains about statistical algorithms that are evaluated and compared on each type of data anomalies. Section 4 gives the conclusion on which statistical algorithm is the best for power consumption anomaly detection.

2 Definitions and Background Study

2.1 Anomaly Detection in Power Consumption

Anomalies are any sort of patterns in data that do not satisfy any notion of well-defined normal data [5]. The anomalies in power consumption data can be any abnormal consumption usage [7]. This can be like a sudden surge or dip in the value of the power consumption, which may occur due to any unwanted circumstances. These unwanted circumstances may occur: when any new devices are being added or removed to the system, the metre is faulty, or even when there is power theft.

2.2 Types of Anomaly

2.2.1 Point Anomaly

These anomalies are the simplest and the most common types that can occur in time series [8, 10]. A point anomaly can be referred to as a point on the data that is distinctly placed from the normal points, more or less like an abrupt outlier as shown in Fig. 1a.

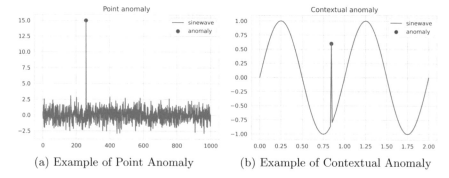

(a) Example of Point Anomaly   (b) Example of Contextual Anomaly

Fig. 1 Types of Anomalies

2.2.2 Contextual Anomaly

The contextual anomalies that occur in a specific context [8, 10]. These anomaly points when considered separately, it might be occurring within the limits compared to the rest of the data points. Hence these are the patterns or points on the time series that deviate from the norm of the expected values in its surroundings, as shown in Fig. 1b.

2.2.3 Collective Anomaly

The collective anomalies are a group of points that are anomalous collectively with respect to spatial and temporal features [8, 10]. These points when taken in isolation may not be an anomaly, but when compared to the rest of the value, these points may appear as an anomaly. For example, when the energy metre produces a null/high value abruptly for few minutes and then gets back to the normal point, it can be considered as a contextual anomaly.

Knowing which type of anomalies are present in the data helps to choose the best combinations of algorithms for anomaly detections. The above-mentioned are the classical types of anomalies that can occur in any time series data. Apart from these, on inspection of the real-world data, several other variations and combinations of the above-mentioned anomalies are found.

2.3 Performance Metrics for Model Evaluations

Performance metrics are the parameters that help to decide how effectively can a model detect the anomaly and non-anomaly points separately [11, 18, 19]. Hence we listed several performance evaluation metrics that are used to measure the performance of an anomaly detection model. Anomaly detection is similar to a binary classification problem. The evaluation parameters can easily explained using a two-class confusion matrix (Table 1).

2.3.1 Sensitivity, Recall, Hit Rate, or True Positive Rate (TPR)

Sensitivity is the proportion of positive classes that were identified correctly to the total positive classes (sum of True Positives (TP) and False Negatives (FN))

$$TPR(Recall) = \frac{TP}{TP + FN} \tag{1}$$

Table 1 Confusion matrix

		True Class	
		Positive	Negative
Predicted Class	Positive	True Positive (TP)	False Negative (FN)
	Negative	False Positive (FP)	True Negative (TN)

2.3.2 Specificity, or True Negative Rate (TNR)

Specificity is the proportion of negative classes that were identified correctly as a negative class to the total negative classes (sum of True Negative (TN) and False Positive (FP))

$$TNR = \frac{TN}{TN + FP} \tag{2}$$

2.3.3 F₁ Score

F_1 score is defined as the weighted average of the Recall and Precision [14]. This Score Defines how effectively a system can retrieve information. It can be used to evaluate the performance of outlier detection models also.

$$F_1\,Score = \frac{2 * Precision * Recall}{Precision + Recall}, \quad where \quad Precision = \frac{TP}{TP + FP} \tag{3}$$

$$F_1\,Score = \frac{2TP}{2TP + FN + FP} \tag{4}$$

2.3.4 Area Under Curve (AUC)

AUC is a most commonly used evaluation parameter for any classification problem. It is also called as Receiver Operating Curve (ROC), This is achieved by plotting the True Positive Rate (TPR) against the False Positive Rate (FPR). A perfect anomaly classifier will have a value one whereas a random classifier will have an AUC value of 0.5.

3　Statistical Algorithms for Univariate Time Series Anomaly Detection

Statistical algorithms include predictive algorithms like ARIMA (Auto-Regressive Integrated Moving Average) [24], Prophet [21] from Facebook, SESD (Seasonal extreme studentized deviate test) [15] from Twitter, and data-oriented models in ADTK (Anomaly Detection Toolkit) [4].

3.1　Library Packages Used

ARIMA was implemented using pmdarima package. Among the several algorithms in ADTK algorithms tested are IQRAD (Inter-Quartile Range Anomaly Detection), outlier anomaly detection, and seasonal anomaly detection. The parametric models of ADTK were hyper-parameter tuned using hyperopt bayesian Optimization [1].

Time series forecasting based anomaly detection using ARIMA and Prophet: ARIMA is classical statistical model for forecasting [2]. Auto-ARIMA function from pmdarima package is utilized to find the parameters of the algorithm automatically. Using these parameters the power consumption data is forecasted for the same input ranges. The mean squared error (MSE) and mean average error (MAE) are calculated on comparing the forecasted and input time series. These errors are called as reconstruction errors. Data points that are beyond a threshold reconstruction error are marked as an anomaly. Hyper optimization along with labelled data is utilized to effectively find this threshold.

Prophet provides extra information in its forecasting results. Apart from the exact forecast point, prophet also provides the dynamic range/band between which the prediction can vary. Any point that is beyond this forecasted dynamic range is considered as an anomaly. These algorithms have an upper hand, since it handles missing data point and public holidays inherently [21].

Anomaly detection using ADTK and SESD: Python Packages ADTK and SESD offer several time series statistical anomaly detection algorithms. Both these algorithms work on basis of the historical data.

(a) IQR AD: This algorithm detect the anomaly on the basis of the inter-quartile range of the historical data. It compares the data with the first and the third quartiles of the historical data. A data point is anomalous when differences are beyond the inter-quartile range (IQR) times a user given parameter c (i.e., IQR*c). For the evaluation purposes factor "c" is tuned using hyper-parameter optimization.

(b) OutlierAD: This class from ADTK package implements classical outlier detection algorithm on the time series data. Scikit-learn outlier detection algorithms like "EllipticEnvelope", "IsolationForest", "LocalOutlierFactor" can be used on a time series data using OutlierAD.

(c) SeasonalAD: This class sequentially uses a pipeline of transformers and anomaly detectors from ADTK package. Transformer is used to obtain residual data after time series decomposition. A data point is identified as anomalous if the residual of seasonal decomposition is anomalously large.

(d) SESD: Seasonal ESD is a statistical algorithm that was developed by Twitter to detect time series anomalies in cloud infrastructures [15]. This algorithm uses seasonal decomposition followed by the use of median and median absolute deviation (MAD)—to effectively detect anomalies. It is tolerant to seasonal spikes in the time series data.

3.2 Results

3.2.1 Dataset

Dataset modelled with twenty-six thousand data points was used to analyse the performance of the algorithms. Eight lakh readings from two hundred energy metres of various ATM location were analysed. The data analysis includes the prepossessing to find power consumption and plotting each sources. On visual and trend analysis, the power consumption was comparatively high on weekdays than on weekends. Thus the custom dataset was modelled as a repeated flattop curves with variable amplitudes. To this anomaly impulses of various specified amplitudes are added or subtracted in the fixed locations. The anomaly points added are tabulated and saved in the dataset. Each of these dataset was divided into train, test, and validation splits, in the ratio 6:2:2. Figure 2 is the plot for the one month custom labelled dataset with 0.5 amplitude anomalies (marked as red). The dataset consists of six sets of data each with twenty-six thousand data points. Each set varies in the amplitude of various anomalies introduced. This approach helps to evaluate amplitude limitations and algorithms.

3.2.2 Evaluation Result Analysis

The results rounded up to 3 decimal places are tabulated, after evaluation. The columns of the result tables signify the peak amplitude of the anomaly in the dataset. The results are tabulated for 0.02, 0.05, 0.1, 0.4, 0.5, 1 units peak amplitudes.

The results of anomaly detection using the forecasting algorithm ARIMA (in Table 2) have its best AUC at 0.603 on dataset with peak 0.1 units. Hence its not effective for the novel model. Even though IQR AD was hyper-tuned using Hyperopt, it was able to achieve only 0.719 AUC for a data of 1unit of peak anomaly. Outlier AD after hyper-tuned was able to achieve a better AUC score for all the anomaly datasets tested but was not able to have an AUC score more than 0.922 (in Table 4) (Table 3).

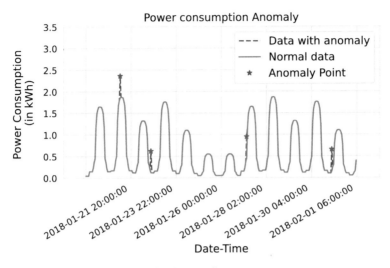

Fig. 2 Labelled data with 0.5 unit amplitude anomalies

Table 2 Results using the AutoArima

AutoArima	P_0.02	P_0.05	P_0.1	P_0.4	P_0.5
F1 Score	0.162	0.178	0.182	0.165	0.163
Recall	0.817	0.914	0.968	0.914	0.914
Precision	0.090	0.098	0.100	0.091	0.089
AUC	0.546	0.589	0.603	0.554	0.548

Table 3 Results using the IQR

IQR	P_0.02	P_0.05	P_0.1	P_0.4	P_0.5	P_1
F1 Score	0.000	0.017	0.025	0.036	0.035	0.051
Recall	0.000	1.000	0.315	0.449	0.438	0.641
Precision		0.008	0.013	0.019	0.018	0.026
AUC	0.500	0.500	0.556	0.624	0.618	0.719

Table 4 Results using the OutlierAD

OutLierAD	P_0.02	P_0.05	P_0.1	P_0.4	P_0.5	P_1
F1 Score	0.213	0.206	0.234	0.182	0.075	0.166
Recall	0.899	0.832	0.944	0.697	0.596	0.584
Precision	0.121	0.117	0.134	0.105	0.040	0.097
AUC	0.922	0.889	0.946	0.823	0.737	0.769

Table 5 Results using the ProphetAD

ProphetAD	P_0.02	P_0.05	P_0.1	P_0.4	P_0.5	P_1
F1 Score	0.014	0.012	0.014	0.036	0.041	0.059
Recall	0.225	0.202	0.225	0.573	0.652	0.809
Precision	0.007	0.006	0.007	0.018	0.021	0.031
AUC	0.478	0.466	0.478	0.656	0.699	0.796

Table 6 Results using the SESD

SESD	P_0.02	P_0.05	P_0.1	P_0.4	P_0.5
F1 Score	0.109	0.108	0.112	0.118	0.111
Recall	0.161	0.161	0.172	0.204	0.204
Precision	0.083	0.081	0.083	0.083	0.077
AUC	0.502	0.500	0.503	0.503	0.494

Table 7 Results using the SeasonalAD

SeasonalAD	P_0.02	P_0.05	P_0.1	P_0.4	P_0.5	P_1
F1 Score	1.000	1.000	1.000	1.000	1.000	0.973
Recall	1.000	1.000	1.000	1.000	1.000	1.000
Precision	1.000	1.000	1.000	1.000	1.000	0.947
AUC	1.000	1.000	1.000	1.000	1.000	1.000

The results of anomaly detection using the Prophet (in Table 5) library and SESD (in Table 6) library were inferior to SeasonalAD. Hence they were not considered for the novel hybrid model.

Among all the statistical algorithms that are tested, hyper-parameter tuned "SeasonalAD" (Table 7) resulted in an AUC score of one, in all the evaluated datasets. Hence, it is assessed to be the best statistical algorithm for the novel hybrid model.

4 Conclusion

In this work, we evaluated all the statistical algorithms on a custom dataset representative of an IoT network with smart metres. We conclude from the published data in this paper that hyper-tuned "SeasonalAD" produced excellent results with highest AUC score. This means, the algorithm was effective in detecting all the non-anomalous and anomalous points accurately on the test data. The experimental study also showed that hyper-parameter tuning was effective in obtaining better results. The analysis of the real world was insightful to make a custom dataset for the purpose of evaluation and training. This study was focused on bench-marking the existing statistical algorithms using a custom modelled dataset.

Our future work includes building the state-of-the-art novel hybrid model developed after the evaluation of statistical and machine learning models [3, 9, 12, 23].

References

1. J. Bergstra, D. Yamins, D.D. Cox, et al., Hyperopt: a python library for optimizing the hyper-parameters of machine learning algorithms, in *Proceedings of the 12th Python in Science Conference*. vol. 13 (Citeseer, 2013), p. 20
2. G.E. Box, G. Jenkins, G. Reinsel, Pages 282–285 time series analysis: forecasting and control (1994)
3. C. Chahla, H. Snoussi, L. Merghem, M. Esseghir, A novel approach for anomaly detection in power consumption data, in *ICPRAM 2019—Proceedings of the 8th International Conference on Pattern Recognition Applications and Methods (Icpram)* (2019). pp. 483–490. https://doi.org/10.5220/0007361704830490
4. Chakraborty, S., Shah, S., Soltani, K., Swigart, A., Yang, L., Buckingham, K.: Building an automated and self-aware anomaly detection system. arXiv preprint arXiv:2011.05047 (2020)
5. V. Chandola, A. Banerjee, V. Kumar, Anomaly detection: a survey. ACM Comput. Surv. (CSUR) **41**(3), 1–58 (2009)
6. N.V. Chawla, K.W. Bowyer, L.O. Hall, W.P. Kegelmeyer, Smote: synthetic minority over-sampling technique. J. Artif. Intell. Res. **16**, 321–357 (2002)
7. J.S. Chou, A.S. Telaga, Real-time detection of anomalous power consumption. Renew. Sustain. Energy Rev. **33**, 400–411 (2014)
8. A.A. Cook, G. Mısırlı, Z. Fan, Anomaly detection for IoT time-series data: a survey. IEEE Internet Things J. **7**(7), 6481–6494 (2019)
9. W. Cui, H. Wang, A new anomaly detection system for school electricity consumption data. Information (Switzerland) **8**(4) (2017). https://doi.org/10.3390/info8040151
10. L. Feng, S. Xu, L. Zhang, J. Wu, J. Zhang, C. Chu, Z. Wang, H. Shi, Anomaly detection for electricity consumption in cloud computing: framework, methods, applications, and challenges. EURASIP J. Wirel. Commun. Netw. **2020**(1), 1–12 (2020)
11. M. Gaur, S. Makonin, I.V. Bajić, A. Majumdar, Performance evaluation of techniques for identifying abnormal energy consumption in buildings. IEEE Access **7**, 62721–62733 (2019)
12. J. Goschenhofer, R. Hvingelby, D. Rügamer, J. Thomas, M. Wagner, B. Bischl, Deep semi-supervised learning for time series classification (2021). arXiv:2102.03622
13. Y. Himeur, A. Alsalemi, F. Bensaali, A. Amira, Anomaly detection of energy consumption in buildings: a review, current trends and new perspectives (2020). arXiv:2010.04560
14. Y. Himeur, A. Alsalemi, F. Bensaali, A. Amira, A novel approach for detecting anomalous energy consumption based on micro-moments and deep neural networks. Cogn. Comput. **12**(6), 1381–1401 (2020)
15. J. Hochenbaum, O.S. Vallis, A. Kejariwal, Automatic anomaly detection in the cloud via statistical learning (2017). arXiv:1704.07706
16. R. Kamal, A.J. Chemmanam, B.A. Jose, S. Mathews, E. Varghese, Construction safety surveillance using machine learning, in *2020 International Symposium on Networks, Computers and Communications (ISNCC)* (IEEE, 2020), pp. 1–6
17. A. Lavin, S. Ahmad, Evaluating real-time anomaly detection algorithms–the numenta anomaly benchmark, in *2015 IEEE 14th International Conference on Machine Learning and Applications (ICMLA)* (IEEE, 2015), pp. 38–44
18. P. Nithin, A. Francis, A.J. Chemmanam, B.A. Jose, J. Mathew, Face tracking robot testbed for performance assessment of machine learning techniques, in *2019 7th International Conference on Smart Computing & Communications (ICSCC)* (IEEE, 2019), pp. 1–5

19. N. Pb, A.J. Chemmanam, B.A. jose, J. mathew, et al., Interactive robotic testbed for performance assessment of machine learning based computer vision techniques. J. Inf. Sci. Eng. **36**(5) (2020)
20. H. Sadeghian, Z. Wang, Autosyngrid: a matlab-based toolkit for automatic generation of synthetic power grids. Int. J. Electr. Power Energy Syst. **118**, 105757 (2020)
21. S.J. Taylor, B. Letham, Forecasting at scale. Am. Stat. **72**(1), 37–45 (2018)
22. X. Wang, T. Zhao, H. Liu, R. He, Power consumption predicting and anomaly detection based on long short-term memory neural network, in *2019 IEEE 4th International Conference on Cloud Computing and Big Data Analytics, ICCCBDA 2019* (2019), pp. 487–491. https://doi.org/10.1109/ICCCBDA.2019.8725704
23. X. Xu, H. Liu, M. Yao, Recent progress of anomaly detection. Complexity **2019** (2019)
24. G.P. Zhang, Time series forecasting using a hybrid arima and neural network model. Neurocomputing **50**, 159–175 (2003)

Author Index

© The Editor(s) (if applicable) and The Author(s), under exclusive license
to Springer Nature Singapore Pte Ltd. 2022
B. Mishra et al. (eds.), *Artificial Intelligence Driven Circuits and Systems*,
Lecture Notes in Electrical Engineering 811,
https://doi.org/10.1007/978-981-16-6940-8

Printed in the United States
by Baker & Taylor Publisher Services